C0-ATC-303

RADIOWAVE PROPAGATION IN SATELLITE COMMUNICATIONS

RADIOWAVE PROPAGATION IN SATELLITE COMMUNICATIONS

Louis J. Ippolito Jr.

VNR VAN NOSTRAND REINHOLD COMPANY
New York

Library of Congress Catalog Card Number: 85-26418
ISBN: 0-442-24011-2

Manufactured in the United States of America

Published by Van Nostrand Reinhold Company Inc.
115 Fifth Avenue
New York, New York 10003

Van Nostrand Reinhold Company Limited
Molly Millars Lane
Wokingham, Berkshire RG11 2PY, England

Van Nostrand Reinhold
480 Latrobe Street
Melbourne, Victoria 3000, Australia

Macmillan of Canada
Division of Gage Publishing Limited
164 Commander Boulevard
Agincourt, Ontario M1S 3C7, Canada

15 14 13 12 11 10 9 8 7 6 5 4 3 2 1

Library of Congress Cataloging-in-Publication Data

Ippolito, Louis J.
Radiowave propagation in satellite communications.

Bibliography: p.
Includes index.
1. Artificial satellites in telecommunication.
2. Radiowave propagation. I. Title.
TK5104.I67 1986 621.3841'1 85-26418
ISBN 0-442-24011-2

PREFACE

Radiowave Propagation in Communications was written with two basic objectives: (1) to present an up-to-date review of the major radiowave propagation phenomena which hinder reliable space communications, and (2) to describe how these propagation phenomena affect the design and performance of satellite communications systems.

Earth-orbiting satellites are employed extensively for the relay of information in a vast array of telecommunications, meteorological, government, and scientific applications. Satellite systems rely on the transmission of radiowaves to and from the satellite and are dependent on the propagation characteristics of the transmission path, primarily the earth's atmosphere. Radiowave propagation thus plays a very important part in the design and ultimate performance of space communications systems.

This book presents, for the first time, the meshing in a single publication of the fundamentals of radiowave propagation factors with a discussion of the practical consequences of these factors on satellite communications systems.

Two major subfields are involved in this book. *Radiowave propagation*, which is basically applied electromagnetic theory, provides the theory and analytical tools for the first several chapters. Later chapters then apply propagation effects to the field of electrical engineering involved with *satellite communications*. The material progresses from the essential aspects of radiowave propagation to the application of practical methods and techniques in the design and performance of satellite communications systems.

This book is written for engineers, scientists, and independent researchers concerned with the specification and performance of satellite communications systems employed in broadcasting, fixed point-to-point transmission, radio navigation, data relay, mobile radio communications, computer communications, and related applications. The recent rapid growth in satellite-based telecommunications (both government and commercial systems) has created a great need for accurate, up-to-date information on radiowave propagation effects on satellite systems design; this book addresses that need for the first time.

The book should have particular appeal for systems engineers interested in specific design information on communications performance while not needing information on the detailed science of the propagation phenomena involved. Step-by-step procedures for the calculation of several key propagation parameters are contained in the book. These procedures include the determination of atmospheric attenuation, rain attenuation prediction (2 methods), rain and ice depolarization, and tropospheric scintillation.

A degree in engineering or the physical sciences is a reasonable prerequisite for a full grasp of this book coverage. A knowledge of communications systems concepts and some familiarity with radiowave propagation principles are also helpful.

This book can be used as a graduate or advanced undergraduate text (or reference) in electrical engineering courses on wave propagation, telecommunications, or space communications.

The subject matter is treated in a conceptual and practical manner, with extensive use of graphics and examples drawn from existing or planned satellite systems. Derivations are introduced only when necessary, and extensive mathematical proofs or manipulations are avoided. More detailed developments are left for appendices, for later reference or further clarification.

Chapter 1 reviews recent developments and trends in space communications which have had an impact on the types of propagation problems important in present space systems.

Chapter 2 presents the fundamentals of radiowave propagation, and discusses the importance of frequency in the types of problems experienced on space communications links.

Chapter 3 describes attenuation by atmospheric gases (oxygen and water vapor), and presents an analytic method for determining atmospheric attenuation on a space link.

Chapter 4 comprehensively reviews the effects of rain and other hydrometeors on space communications. The classical development model for rain attenuation is presented, along with recent direct measurements observed from orbiting satellites. The effects of rain are recognized as the major problem in space communications at frequencies above 10 GHz.

Chapter 5 reviews rain attenuation models developed over the past two decades for the prediction of rain effects on satellite links. The complete step-by-step calculation procedures for two of the better models (Global Model and CCIR Model) are presented in Appendices D and E.

Chapter 6 describes depolarization effects caused by rain, ice, and multipath. The latest prediction models are presented and reviewed.

Chapter 7 describes the effects of radio noise on satellite communications. Noise from atmospheric gases, clouds, rain, surface emissions, and extra-terrestrial sources is covered.

Chapter 8 covers several propagation factors often considered to be "secondary" problems, but which can be important in specific applications. These factors are scintillation, bandwidth coherence, antenna gain degradation, and angle of arrival effects.

Chapter 9 provides a quantitative description of propagation effects on communications satellite link performance. Link performance is analyzed step-by-step through the system, and the effects of path attenuation and noise are vividly described by graphical presentations.

Chapter 10 introduces, again for the first time in a single text, the new area of restoration techniques for overcoming severe attenuation. Methods such as site diversity, power control, spot beams, and dynamic forward error correction are described and evaluation methods reviewed.

Chapters 2 through 8 focus, for the most part, on the basic propagation factors important to present-day space communications, while Chapters 9 and 10 describe the practical consequences of these factors on satellite communications systems and applications.

Examples and case studies are provided throughout the book to further a real understanding of the effects of propagation on space communications systems.

I would like to acknowledge the contributions of the many organizations and individuals whose work and efforts are referenced in this book. I have had the privilege of knowing many of the researchers personally through my affiliations with NASA, the CCIR, and SPACECOM. It was through extensive discussions and interchange of ideas that enabled this book to be developed, and I thank all the foregoing, too numerous to be mentioned individually, for such indispensable help and guidance.

Finally, I gratefully acknowledge the support and encouragement of my wife, Sandi, whose unlimited patience I could always count on during the long hours of preparation and writing. This book is dedicated to Sandi, and to our children, Karen, Rusty, Teddy, and Cathie.

LOUIS J. IPPOLITO
Columbia, Maryland

CONTENTS

RADIOWAVE PROPAGATION IN SATELLITE COMMUNICATIONS

CHAPTER 1
INTRODUCTION

Earth-orbiting satellites are employed extensively for the relay of information in a vast array of telecommunications, meteorological, and scientific applications. These satellite systems rely on the transmission of radiowaves to and from the satellite and are dependent on the propagation characteristics of the transmission path, primarily the earth's atmosphere. Radiowave propagation thus plays an important part in the design and ultimate performance of space communications systems.

1.1. PURPOSE AND OBJECTIVES

This book provides a concise review of the effects of the earth's atmosphere on space communications. Techniques used in the determination of atmospheric effects, as influenced by frequency of operation, elevation angle to the satellite, type of modulation, and other system factors are discussed. Recent data on performance characteristics obtained from direct measurements on satellite links operating at frequencies to above 30 GHz (Gigahertz) are included.

The objectives of this book are: (a) to present an up-to-date review of the major radiowave propagation phenomena which hinder reliable space communications, and (b) to describe how these propagation phenomena affect the design and performance of satellite communications systems.

The book is written for those concerned with the design and performance of satellite communications systems employed in broadcasting, fixed point-to-point, radio-navigation, data-relay, computer communications, and related applications. The recent rapid growth in satellite communications has created a need for accurate information on radiowave propagation effects on satellite systems design, and this book addresses that need for the first time.

The first part of the book introduces and develops the fundamental radiowave propagation effects which are important in modern space communications systems. The second part then provides the analytical techniques and practical methods necessary to evaluate the design and performance of specific satellite telecommunications systems and applications.

The relative importance of radiowave propagation factors in space communications depends to a large extent on the frequency of operation of the earth-space links, as well as on the local climatology, local geography, type of transmission, and the elevation angle to the satellite. Generally, the effects become more significant as the frequency increases and as the elevation angle decreases; however, there are exceptions, which will be pointed out in the appropriate chapters. The random nature and general unpredictability of the phenomena which produce the propagation effects add further dimensions of complexity and uncertainty in the evaluation of radiowave propagation in space communications. Consequently, statistical analyses and techniques are useful for evaluation of propagation effects on communication links.

Even apparent "clear sky" conditions can produce propagation effects which can degrade or change the transmitted radiowave. Gases present in the earth's atmosphere, particularly oxygen and water vapor, interact with the radiowave and reduce the signal amplitude by an absorption process. Turbulence or rapid temperature variations in the transmission path can cause amplitude and phase scintillation or depolarize the wave. Clouds, fog, dirt, sand, and even severe air pollution can cause observable propagation effects. Also, background "sky noise" will always be present, and contributes directly to the noise performance of the communications receiver system.

The degrading effect of precipitation in the transmission path is a major concern associated with space communications systems, particularly for those systems operating at frequencies above 10 GHz. At those frequencies, absorption and scattering caused by rain, hail, ice crystals, or wet snow, can cause a reduction in transmitted signal amplitude (attenuation), which can reduce the reliability and/or performance of the communication link. Other effects can be generated by precipitation on the earth-space path. They include depolarization, rapid amplitude and phase fluctuations (scintillation), antenna gain degradation, and bandwidth coherence reduction.

Before we begin the discussion of propagation effects, a brief review of the historical development of space communications will be helpful in pointing out some of the trends which have had a direct impact on the types of propagation phenomena that are important in space communications.

1.2. DEVELOPMENTS AND TRENDS IN SPACE COMMUNICATIONS

The growth in the number of orbiting satellites launched for space science and applications over the past three decades has been phenomenal. Prior to 1970, only a handful of communications satellites had been placed in active geostationary orbit. These included seven INTELSAT international communications satellites, NASA experimental satellites SYNCOM-3, ATS-1,3, and 5, the first of the United Kingdom's SKYNET satellites, and the first TACSAT of the U.S. Air Force [1.1, 1.2]. The interval 1970–1974 saw over two dozen more geostationary satellites placed in orbit, including communications satellites of Canada and the U.S.S.R. The period 1975–1979 produced over seventy new satellites in orbit, including several which introduced new services in the meteorological, mobile, and direct broadcasting areas. Japan, Indonesia, and the European Space Agency also joined the list of satellite producing countries, while the U.S. domestic communication satellite fleet expanded with systems developed by COMSAT General, RCA Americom, and Western Union Telegraph Company. The 1980s have produced an explosion of new satellites, services, and capabilities in space, with hundreds of new satellites planned for launch well into the 1990s. France, Italy, the People's Republic of China, India, and Australia have been or soon will be added to the list of satellite producing nations, and the emergence of second generation satellites and regional systems has further accelerated the growth of satellites in earth orbit.

There are three major trends in the development of satellite communications which have increased the need for a further awareness and understanding of the propagation factors involved in space communications. First, the rapid increase in the utilization of the frequency spectrum in the bands below 6 GHz which are allocated for space services has produced conditions of spectrum crowding and frequent sharing of the spectrum. These problems have required the systems designer to look to the higher frequency bands to relieve the congestion. The bands allocated for fixed satellite service at 14/12 GHz and for broadcasting satellite service at 17/12 GHz are already being utilized on present satellites or are planned for operational systems over the next few years.* NASA's Tracking and Data Relay Satellite, TDRS, operates in the 15 GHz band. Japan has operated satellites in the 30/20 GHz band for space research and communications experiments. The United States and Italy are developing advanced multiple beam satellites that will operate in the 30/20 GHz bands for high data rate communications applications. Allocations for many other space services, including space research, earth-exploration, meteorological, mobile, and radio navigation, are available from 10 GHz to well over 100 GHz.

The trend to higher and higher operational frequencies can be observed from Figure 1-1, which presents the number of geostationary satellites in each of several frequency bands launched into orbit or planned for launch, for five year intervals beginning in 1965 through the mid-1980s. The greatest expansion is seen in the 11-17 GHz region, where several operational satellites in the fixed satellite service and the broadcasting service are planned for implementation.

A second major trend in satellite communications that has tended to push the

*Following the usual convention, satellite frequency bands are specified by nominal uplink frequency/nominal downlink frequency designations.

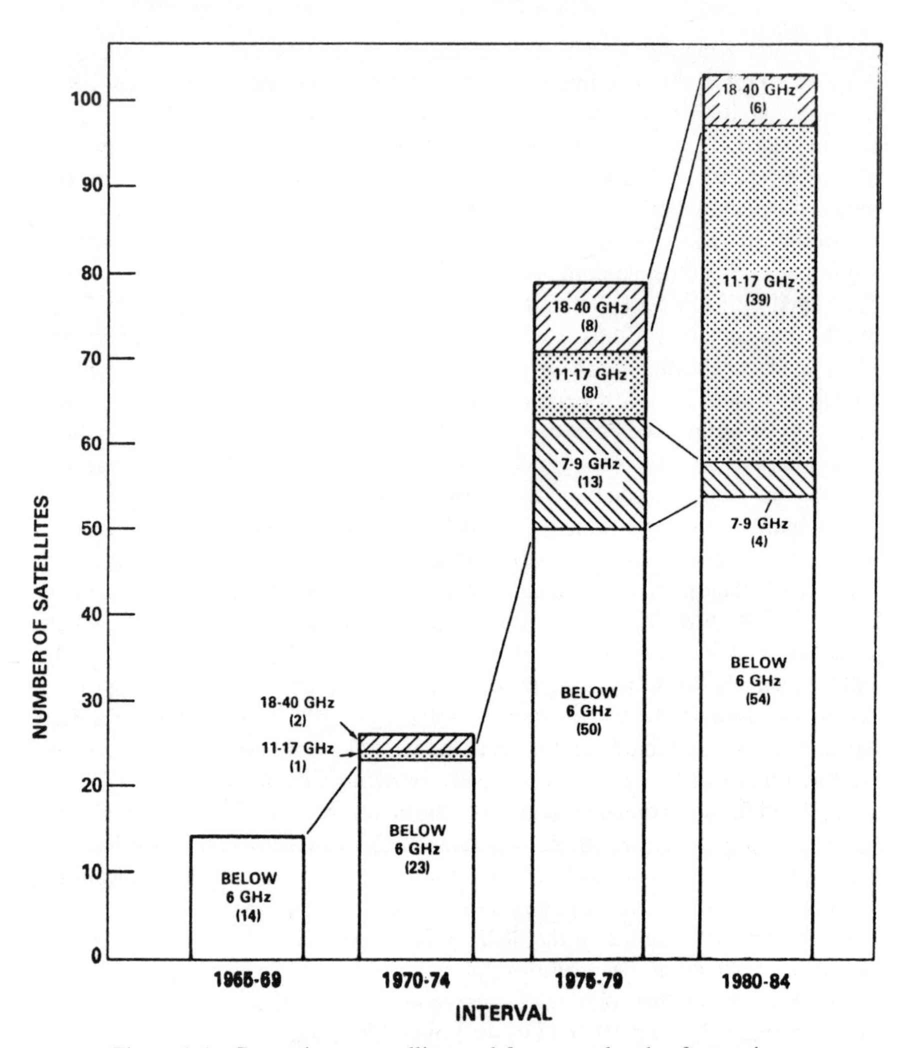

Figure 1-1. Geostationary satellites and frequency bands of operation.

development of satellite systems to higher operating frequencies has been the expanded need for more bandwidth to accommodate the exploding information transfer requirements of our society. Present fixed satellite service allocations in the 6/4 GHz and 8/7 GHz frequency bands, which support the majority of present domestic, international, and military communications, totals 500 MHz in each of the two bands. Above 10 GHz, however, the available bandwidth increases drastically. Present allocations adopted at the 1979 World Administrative Radio Conference (WARC) provide for bandwidths of 1000 MHz in the

6/4 GHz region, 1000 MHz in the 14/12 GHz region, 3.5 GHz in the 30/20 GHz region, and 3 GHz in the 43/40 GHz region. The total bandwidth presently allocated to fixed satellite service in the bands above 10 GHz is over 80 GHz, extending to the upper limit of allocations at 275 GHz.

A third factor in the move to higher operating frequencies concerns the limited number of orbital positions or "slots" available in the geostationary orbit. Satellites which operate at the same frequencies must be separated in orbit to avoid interference. The Federal Communications Commission (FCC) has the responsibility for approving satellite orbital locations and for determining the policy for domestic United States orbit spacing. Until mid-1983, the FCC required a spacing of 4 degrees for domestic fixed service satellites in the 6/4 GHz band, and a spacing of 3 degrees for the 14/12 GHz band. (The spacing for Canada in the 6/4 GHz band is 5 degrees, by bilateral Canadian–U.S. agreement.) The FCC spacing plan approved in 1983 called for a phased reduction in orbital spacing which would result in 2 degree spacing for both the 6/4 GHz and 14/12 GHz bands by 1987.

The segment of the geostationary arc permitting the use of ground stations with elevation angles of 5 degrees or higher in the contiguous 48 United States ranges from about 60 to 140 degrees West Longitude. With a nominal spacing between fixed service communications satellites of 4 degrees, 21 slots are available. For a spacing of 2 degrees, the total is 41 slots. If coverage is to be provided to all 50 states, the usable segment of the arc reduces to 100–140 degrees, and the number of slots at the two separations are 11 and 21, respectively. Through 1983 domestic United States and Canadian organizations had 18 slots at 6/4 GHz and 6 slots at 14/12 GHz occupied or committed for fixed satellite service. Launches through 1984 raise the number of slots occupied to 23 at 6/4 GHz and 12 at 14/12 GHz. Through the mid-1980s the number of slots occupied increases to 33 at 6/4 GHz and to 19 at 14/12 GHz, including two satellites to be launched by Mexico which operate in both frequency bands. Thus, even in the absence of other Western Hemisphere nations employing space systems, and not accounting for the shared 17/12 allocation with broadcasting service satellites, which would require even wider spacing, the American arc could be essentially filled at 6/4 GHz and near 50% of capacity at 14/12 GHz within the next few years.

A number of techniques have been proposed or are under active development for a more efficient utilization of the orbital arc. These techniques include frequency reuse employing polarization diversity, site diversity, spot beams for increased reliability during adverse weather conditions, multiple beam switched or scanned satellite antennas to cover large service areas, and larger aperture narrow beamwidth systems to reduce intersatellite interference. All of these techniques, particularly those requiring larger aperture antennas and multiple beam switching on the satellite, can be most efficiently accomplished at higher frequencies, where component size and weight are more compatible with spaceborne constraints.

With the trends that have developed in space communications, i.e., the move to higher frequencies, wider bandwidth systems, and more advanced frequency and orbit reuse techniques, the impact of radiowave propagation on overall system design and performance cannot be understated.

1.3. FREQUENCY ALLOCATIONS AND REGULATORY ASPECTS

An introduction to satellite communications would not be complete without acknowledgment of the role that frequency allocations and regulatory considerations play in systems design and performance. The frequency of operation is perhaps the major determining factor in the propagation performance of a satellite communications link. The systems designer must operate within the constraints of international and domestic regulations related to choice of frequency, allowed power radiation, orbit location, and many other factors which are often predetermined or limited by in-place regulations.

The Federal Communications Commission (FCC) provides the regulatory function for all domestic telecommunications systems in the United States, except for Federal Government systems. The National Telecommunications and Information Administration (NTIA) provides this function for Federal Government telecommunications.

International telecommunications are governed by treaty documents of the International Telecommunications Union (ITU). The ITU was formed in 1932 from the International Telegraph Union, which originated in 1865. The ITU is a specialized agency of the United Nations, with over 150 current member nations (administrations).

The radio regulations which serve as the basis for international telecommunications are developed at periodic Administrative Radio Conferences. The 1979 World Administrative Radio Conference, WARC-79, for example, specified frequency allocation tables for telecommunications services up to 275 GHz. Technical information that supports the Administrative Conferences is provided by the ITU's International Radio Consultative Committee (CCIR) at its quadriennial Plenary Assemblies. The Plenary-approved documents, published in three languages, are available in the United States (in English) through the U.S. Department of Commerce's National Technical Information Service in Springfield, Virginia.

The CCIR generates technical reports and recommendations through thirteen study groups (SG) of experts from member administrations. Two of the study groups are directly concerned with radiowave propagation: SG 5—Propagation in Non-ionized Media, and SG 6—Ionospheric Propagation. Other study groups involved with satellite communications include: SG 4—Fixed Services Using

Communications Satellites, and SG 10/11—Broadcasting Services (Sound/Television). CCIR reports and documents provide an excellent source of the latest information on the state of global telecommunications development, and they are referenced quite extensively in this book.

1.3.1. Frequency Allocations for Satellite Communications

International frequency allocations specified by the ITU are based on the type of service provided by the telecommunications system. The allocations include terrestrial, satellite, and inter-satellite delivered services. Table 1-1 lists the satellite services as designated by the ITU for the purpose of frequency allocation. The designations are very specific, and several are subdivided into additional services. For example, mobile satellite service is one designation, as are aeronautical mobile, land mobile, and maritime mobile.

The fixed satellite service refers to point-to-point communications links where the *ground terminals* remain fixed in place. Mobile satellite service covers point-to-point communications links where one or both ends of the link are moving during transmissions. Broadcasting satellite service refers to single-point-to-multiple-points links where a program is radiated over a "service area" to many receive-only terminals.

The ITU divides the surface area of the earth into three Regions for the purpose of frequency allocation. The three Regions are shown on the map of Figure 1-2. Allocations are generally assigned independently in each Region, subject

Table 1-1. Satellite Services Designated by the International Telecommunications Union (ITU).

- Aeronautical mobile satellite
- Aeronautical radionavigation satellite
- Amateur satellite
- Broadcasting satellite
- Earth-exploration satellite
- Fixed satellite
- Inter-satellite
- Land mobile
- Maritime mobile satellite
- Maritime radionavigation satellite
- Meterological satellite
- Mobile satellite
- Radionavigation satellite
- Space operations
- Space research
- Standard frequency satellite

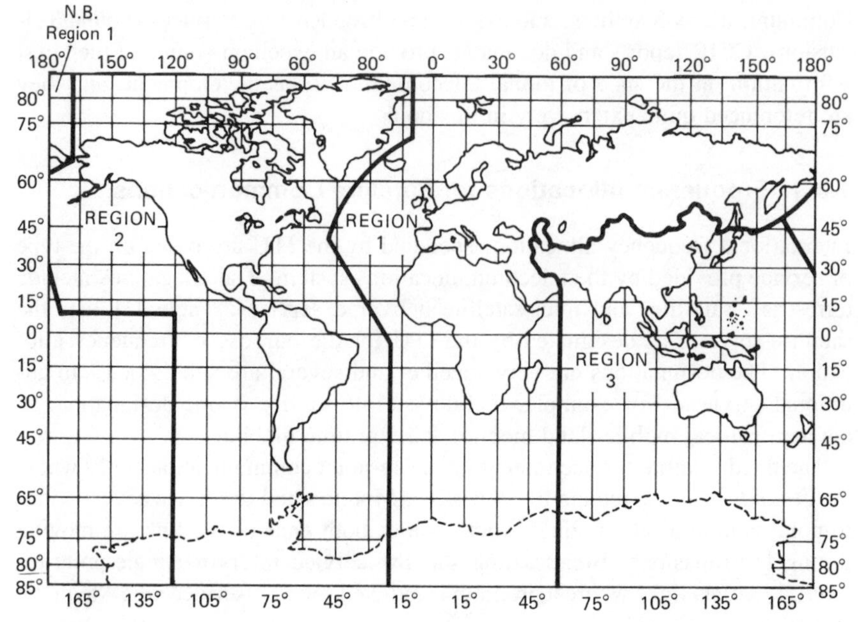

Figure 1-2. Telecommunications service regions, as designated by the ITU.

to aggreement of the administration in that Region, although many allocations are the same for all three Regions. Interference criteria between Regions can be a significant factor in frequency allocations, particularily near the borders of adjacent regions.

Table 1-2 lists the fixed satellite service (FSS) international frequency allocations for Region 2 as specified by the 1979 World Administrative Radio Conference (WARC-79). Also shown on the table is the bandwidth allocated at each frequency band.

Most domestic communications satellites presently operate in the 5.85–6.65 GHz uplink/3.4–4.2 GHz downlink band (C-band). The 14.0–14.5 GHz uplink/ 11.7–12.2 GHz downlink band (Ku-band), however, is rapidly becoming utilized as the available C-band orbital locations are filled up with operating satellites. The 27–31 GHz uplink/17.7–21.2 GHz downlink band (Ka-band) is the next higher frequency band of interest. It is being considered for domestic satellite communications by several nations as the Ku-band becomes fully utilized. Several satellites, such as ANIK-B, the INTELSAT V series, and the SPACENET series, are "hybrid" satellites, that is they operate at both C-band and Ku-band.

Tables 1-3, a and b, list mobile satellite service (MSS) allocations for Region 2, below 3 GHz and above 3 GHz respectively. Those frequencies shown in

Table 1-2. Frequency Allocations in Region 2 for Fixed Satellite Service.

Band (GHz)	Link	Bandwidth (MHz)
2.5–2.69	D	190
3.4–4.2	D	800
4.5–4.8	D	300
5.85–7.075	U	1225
7.25–7.75	D	500
7.9–8.4	U	500
10.7–11.7	D	1000
11.7–12.2[a]	D	500
12.7–13.25	U	550
14–14.5	U	500
17.7–21.2	D	3500
27–31	U	4000
37.5–40.5	D	3000
42.5–43.5	U	1000
50.4–51.4	U	1000
71–75.5	U	4500
81–84	D	3000
92–95	U	4000
102–105	D	3000
149–164	D	15000
202–217	U	15000
231–241	D	10000
265–275	U	10000

U—uplink; D—downlink.
[a]As modified by RARC-83

parentheses are ''secondary allocations,'' (i.e., another service has the primary allocation to operate in the band). The MARISAT maritime mobile satellites operate in the 1.6 GHz uplink/1.5 GHz downlink bands. Land mobile satellites operating in the 806 to 890 MHz band are under consideration in the United States and Canada.

Table 1-4 lists the broadcasting satellite service (BSS) frequency allocations for Region 2. BSS systems being developed in the United States will operate in the 12.2 to 12.7 GHz downlink band, with feeder links (uplinks) in the 17.3 to 18.1 GHz band. Some interim BSS systems operate in the 11.7 to 12.2 GHz FSS downlink band using on-orbit domestic satellites such as SBS and ANIK.

It should be noted that the frequency allocation tables are heavily ''footnoted'' with special constraints and/or conditions of operation. Many of the bands are shared by several services, both satellite and terrestrial based. Reference should be made to the *Manual of Regulations and Procedures for Federal Radio Frequency Management*, published by the NTIA [1.3] for a full

Table 1-3(a). Frequency Allocations in Region 2 for Mobile Satellite Service (below 3 GHz)

Band (MHz)	Link	Service
(117.98–137)	U, D	A
121.45–121.55	U	L, M, A (D/S)
235.4–399.9	U	L, M, A
242.95–243.05	U	L, M, A (D/S)
273–322	U, D	L, M, A
405.5–406[a]	U	L, M
406–406.1	U	L, M, A
406.1–410[a]	U	L, M
(608–614)	U	L, M
806–890	U, D	L, M
1530–1544	D	M
1544–1545	D	L, M, A (D/S)
1545–1559	D	A
1610–1626.5	U, D	A
1626.5–1645.5	U	M
1645.5–1646.5	U	L, M, A (D/S)
1646.5–1660.5	U	A

U—uplink; D—downlink.
L—land mobile; M—maritime mobile; A—aeronautical mobile.
D/S—distress and safety only.
( )—secondary allocation.
[a]Canada only.

Table 1-3(b). Frequency Allocations in Region 2 for Mobile Satellite Service (above 3 GHz).

Band (GHz)	Link	Service
5–5.25	U, D	A
7.25–7.375	D	L, M, A
7.9–8.025	U	L, M, A
(14–14.5)	U	L
15.4–15.7	U, D	A
(19.7–21.2)	D	L, M, A
29.5–31	U	L, M, A
39.5–40.5	D	L, M, A
43.5–47	U, D	L, M, A
(50.4–51.4)	U	L, M, A
66–71	U, D	L, M, A
71–74	U	L, M, A
81–84	D	L, M, A
95–100	U, D	L, M, A
134–142	U, D	L, M, A
190–200	U, D	L, M, A
252–265	U, D	L, M, A

U—uplink; D—downlink.
L—land mobile; M—maritime mobile; A—aeronautical mobile.
( )—secondary allocation.

Table 1-4. Frequency Allocations in Region 2 for Broadcasting Satellite Service

Band (GHz)	Link	Bandwidth (MHz)
0.620–0.790[a]	D	170
2.5–2.7	D	200
12.2–12.7[b]	D	500
14.5–14.8	U	300
17.3–18.1	U	800
40.5–42.5	D	2000
47.2–50.2[c]	U	3000
84–86	D	2000

[a]For FMTV, with power flux density limitation.
[b]As modified by RARC-83.
[c]Limited to feeder links for 40.5–42.5 GHz band.

description of the frequency allocation tables for all of the servies. This document lists both international and domestic frequency allocations, which can be different for particular frequency bands and services.

REFERENCES

1.1. Morgan, W. L., "Satellite Utilization of the Geostationary Orbit," *COMSAT Technical Review*, Vol. 6, No. 1, Spring 1976, pp. 195–205.
1.2. Morgan, W. L., "Geosynchronous Satellite Log," *COMSAT Technical Review*, Vol. 8, No. 1, Spring 1978, pp. 219–237.
1.3. *Manual of Regulations and Procedures for Federal Radio Frequency Management*, Cat. No. Pr ex 18.8:R11, Superintendent of Documents, Washington, D.C. 20402.

CHAPTER 2
FUNDAMENTALS OF RADIOWAVE PROPAGATION

In this chapter the fundamental elements of radiowave propagation are introduced. The basic transmission elements, such as free space attenuation, antenna gain, and polarization, are defined. Radiowave propagation modes and propagation mechanisms are discussed and the importance of the ionosphere and the troposphere in space communications is highlighted. Finally, the major radiowave propagation factors in the frequency bands of interest to space communications are introduced and described.

2.1. TRANSMISSION PRINCIPLES

An electromagnetic wave, referred to as a radiowave at radio frequencies, is characterized by variations of its electric and magnetic fields. The oscillating motion of the field intensities vibrating at a particular point in space at a frequency f excites similar vibrations at neighboring points, and the radiowave is said to travel or to *propagate*. The wavelength λ of the radiowave is the spatial separation of two successive oscillations, which is the distance the wave travels during one cycle of oscillation. The frequency and wavelength in free space are related by

$$\lambda = \frac{c}{f} \qquad (2\text{-}1)$$

where c is the phase velocity of light in a vacuum.

Consider a radiowave propagating in free space from a point source P of power p_t watts. The wave is isotropic in space, i.e., spherically radiating from the point P, as shown in Figure 2-1(a). The power flux density (or power density), over the surface of a sphere of radius r_a from the point P is given by

$$(pfd)_A = \frac{p_t}{4\pi\, r_a^2}, \quad \text{watts/m}^2 \qquad (2\text{-}2)$$

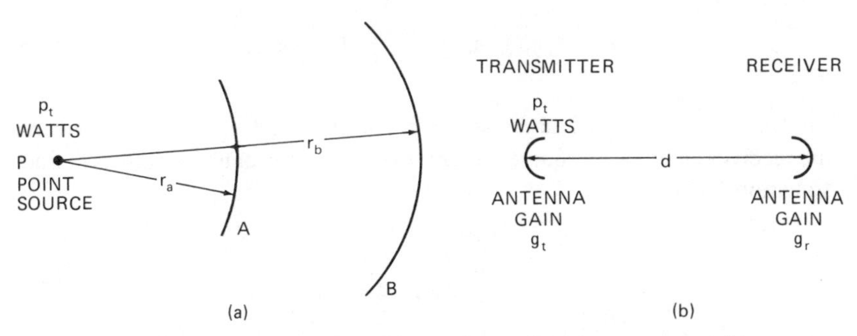

Figure 2-1. Radiowave propagation for (a) an isotropic point source and (b) a communications link with directive antennas.

Similarly, at the surface B, the density over a sphere of radius r_b is given by

$$(pfd)_B = \frac{p_t}{4\pi\, r_b^2}, \quad \text{watts/m}^2 \tag{2-3}$$

The ratio of power densities is given by

$$\frac{(pfd)_A}{(pfd)_B} = \frac{r_b^2}{r_a^2} \tag{2-4}$$

where $(pfd)_B$ is less than $(pfd)_A$. This relationship demonstrates the well known *inverse square law* of radiation; the power density of a radiowave propagating from a source is inversely proportional to the square of the distance from the source.

2.2. ANTENNA GAIN AND FREE SPACE ATTENUATION

Physically realizable transmitters and receivers used in communications systems are not isotropic, but will have some directivity in the form of an antenna with gain. The *gain* of an antenna is defined as

$$g = \frac{4\pi\, A_e}{\lambda^2} \tag{2-5}$$

where A_e is the effective area of the antenna, which is somewhat less than its physical aperture area due to ohmic and aperture losses.

Consider now a receiver with an antenna of gain g_r located a distance d from a transmitter of p_t watts and antenna gain g_t, as shown in Figure 2-1(b).

The power p_r intercepted by the receiving antenna will be

$$p_r = (pfd)_r A_r = \frac{p_t g_t}{4\pi d^2} A_r, \quad \text{watts} \tag{2-6}$$

where $(pfd)_r$ is the power flux density at the receiver and A_r is the effective area of the receiver antenna, in square meters. In terms of the antenna gain as defined in Equation (2-5),

$$p_r = \frac{p_t g_t}{4\pi d^2} \frac{g_r \lambda^2}{4\pi} \tag{2-7}$$

Rearranging terms gives

$$p_r = p_t g_t g_r \left(\frac{\lambda}{4\pi d}\right)^2 \tag{2-8}$$

where the term in brackets is defined as the *free space path loss*, l, i.e.

$$l = \left(\frac{\lambda}{4\pi d}\right)^2 \tag{2-9}$$

The free space path loss, or free space attenuation, is often expressed in decibels, dB, as a positive quantity, i.e.,

$$L(\text{dB}) = 20 \log \left(\frac{4\pi d}{\lambda}\right) \tag{2-10}$$

Free space attenuation is present for all radiowaves propagating in free space or in regions whose characteristics approximate the uniformity of free space, such as the earth's atmosphere.

Figure 2-2 shows the path loss as a function of frequency for several representative path lengths, including the distance to a synchronous satellite, for ground station elevation angles of 0 and 90 degrees.

The gain of a circular parabolic reflector antenna can be expressed in dBi, decibels above isotropic, as

$$G(\text{dB}) = 10 \log \left(\eta \frac{\pi^2 D^2}{\lambda^2}\right) \tag{2-11}$$

where η is the antenna efficiency, D is the antenna diameter, and λ is the wavelength.

Figure 2-3 presents antenna gain as a function of diameter for several rep-

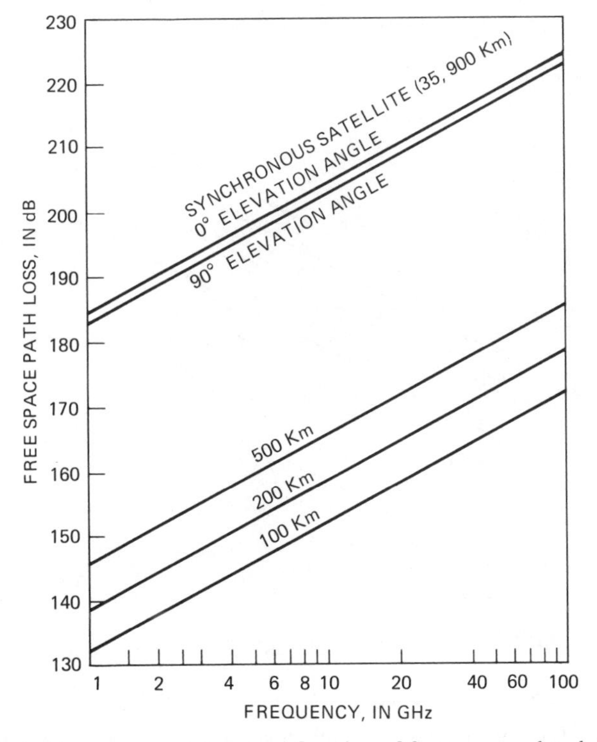

Figure 2-2. Free space path loss as a function of frequency and path length.

resentative frequencies used in satellite communications. An antenna efficiency of 55% is assumed for the calculations.

2.3. POLARIZATION

The polarization of the radiowave is determined by the orientation of the electric and magnetic field vectors at a fixed point in space. A *linearly polarized wave* is a wave whose electric and magnetic field vectors always lie along fixed directions at a point in space as a function of time. The direction of the electric field vector determines the sense of the linear polarization, i.e., horizontal, vertical, or at a specified angle, with respect to a local reference.

A *circularly polarized wave* is a wave whose electric and magnetic field vectors rotate at the rate of the wave frequency and describe a circle at a fixed point, as a function of time. The sense of the circular polarization, either clockwise (right-handed), or counterclockwise (left-handed), is determined by the direction of rotation of the electric field vector as seen by an observer looking in the direction of travel of the propagating wave.

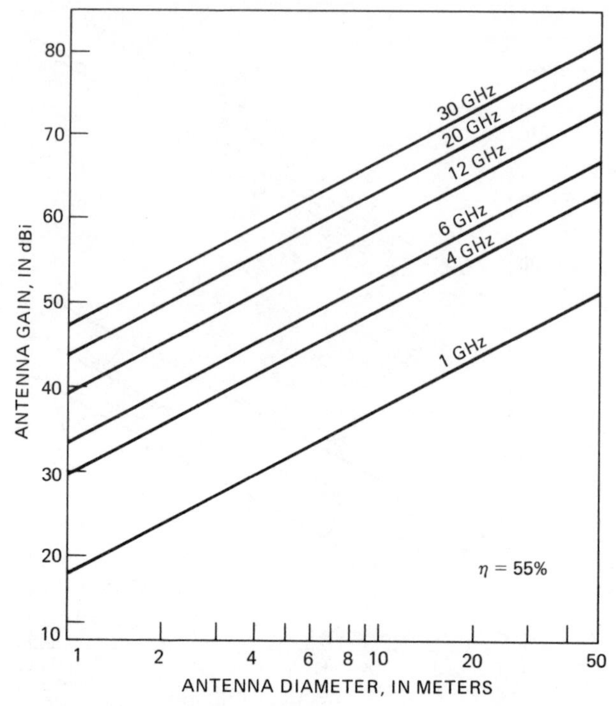

Figure 2-3. Gain of a circular parabolic reflector antenna. Efficiency equals 55%.

An *elliptically polarized wave* is a wave whose electric and magnetic field vectors rotate at the rate of the wave frequency and describe an ellipse at a fixed point, as a function of time. The sense of the elliptical polarization is defined in the same way as for circular polarization.

2.4. RADIOWAVE FREQUENCY AND SPACE COMMUNICATIONS

The frequency of the radiowave is a critical factor in determining whether impairments to space communications will be introduced by the Earth's atmosphere. A radiowave will propagate from the Earth's surface to outer space provided its frequency is high enough to penetrate the ionosphere, which is the ionized region extending from about 50 km to roughly 2000 km above the surface. The various regions (or layers) in the ionosphere, designated *D*, *E*, and *F*, in order of increasing altitude, act as reflectors or absorbers to radiowaves at frequencies below about 30 MHz, and space communications is not possible. As the frequency is increased, the reflection properties of the *E* and *F* layers are reduced and the signal will propagate through. Radiowaves above about 30 MHz will propagate through the ionosphere, however, the properties of the wave could be modified or degraded to varying degrees depending on frequency, geographic location, and time of day. Ionospheric effects tend to become less significant as the frequency of the wave increases, and above about 3 GHz the ionosphere is essentially transparent to space communications, with some exceptions which will be discussed later.

Several types of radiowave propagation modes can be generated in the Earth's atmosphere, depending on transmission frequency and other factors. Below the ionospheric penetration frequency, a radiowave will propagate along the Earth's surface, as illustrated in Figure 2-4(a). This mode is called *ground wave propagation*, and consists of three components: a direct wave, a ground reflected wave, and a surface wave which is guided along the Earth's surface. This mode supports broadcasting and communications services, such as the AM broadcast band, amateur radio, radionavigation, and land mobile services.

A second type of terrestrial propagation mode, called an *ionospheric* or *sky wave*, can also be supported under certain ionospheric conditions, as shown in Figure 2-4(b). In this mode, which occurs at frequencies below about 300 MHz, the wave propagates toward and returns from the ionosphere, hopping along the surface of the Earth. This frequency range includes the commercial FM and VHF-television bands as well as aeronautical and marine mobile services.

Above about 30 MHz, and up to about 3 GHz, reliable long distance over-the-horizon communications can be generated by a scattering of energy from refractive index irregularities in the troposphere, the region from the Earth's surface up to about 10–20 km in altitude. This mode of propagation, termed a *tropospheric* or *forward scattered wave* [see Figure 2-4(c)], is highly variable

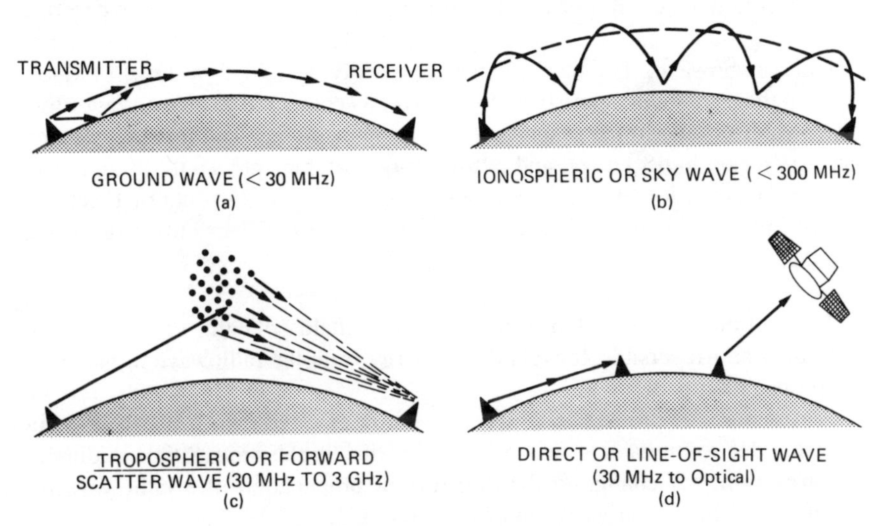

Figure 2-4. Radiowave propagation modes.

and is subject to intense fluctuations and interruptions. This mode has been and is being used, however, for long distance communications when no other means are available. Tropospheric scatter propagation can also be a factor in space communications when the scattered signal from a ground transmitter interferes with a ground receiver operating in the same frequency band. The scattered signal will appear as noise in the receiver and can directly contribute to a degradation in system performance.

Finally, at frequencies well above the ionospheric penetration frequency, *direct* or *line-of-sight propagation* predominates, and this is the primary mode of operation for space communcations [see Figure 2-4(d)]. Terrestrial radio-relay communications and broadcasting services also operate in this mode, often sharing the same frequency bands as the space services.

Line-of-sight space communications will proceed unimpeded as the frequency of transmission is increased up to frequencies where the gaseous constituents of the troposphere, primarily oxygen and water vapor, will absorb energy from the radiowave. At certain specific ''absorption bands,'' where the radiowave and gaseous interaction are particularly intense, space communications are severely limited. It is in the ''atmospheric windows'' between absorption bands that practical earth–space communications have developed, and it is in these windows that we will focus our attention in our study of radiowave propagation factors.

2.5. RADIOWAVE PROPAGATION MECHANISMS

Before beginning our detailed discussion of radiowave propagation in space communications, it will be useful to introduce the general terms used to describe the propagation phenomena, or mechanisms, which can affect the characteristics of a radiowave. The mechanisms are usually described in terms of variations in the signal characteristics of the wave, as compared to the natural or free space values found in the absence of the mechanism. The definitions presented here are meant to be general and introductory. Later chapters will discuss them in more detail. Most of the definitions are based on the Institute of Electrical and Electronics Engineers (IEEE) Standard Definitions of Terms for Radio Wave Propagation [2.1].

Absorption. A reduction in the amplitude (field strength) of a radiowave caused by an irreversible conversion of energy from the radiowave to matter in the propagation path.

Scattering. A process in which the energy of a radiowave is dispersed in direction due to interaction with inhomogeneities in the propagation medium.

Refraction. A change in the direction of propagation of a radiowave resulting from the spatial variation of refractive index of the medium.

Diffraction. A change in the direction of propagation of a radiowave resulting from the presence of an obstacle, a restricted aperture, or other object in the medium.

Multipath. The propagation condition that results in a transmitted radiowave reaching the receiving antenna by two or more propagation paths. Multipath can result from refractive index irregularities in the troposphere or ionosphere, or from structural and terrain scattering on the Earth's surface.

Scintillation. Rapid fluctuations of the amplitude and the phase of a radiowave caused by small-scale irregularities in the transmission path (or paths) with time.

Fading. The variation of the amplitude (field strength) of a radiowave caused by changes in the transmission path (or paths) with time. The terms fading and scintillation are often used interchangeably; however, fading is usually used to describe slower time variations, on the order of seconds or minutes, while scintillation refers to more rapid variations, on the order of fractions of a second in duration.

Frequency Dispersion. A change in the frequency and phase components across the bandwidth of a radiowave, caused by a dispersive medium. A dispersive medium is one whose constitutive components (permittivity, permeability, and conductivity) depend on frequency (temporal dispersion) or wave direction (spatial dispersion).*

Many of the mechanisms described above can be present on the transmission path at the same time and it is usually extremely difficult to identify the mechanism or mechanisms which produce a change in the characteristics of the transmitted signal. This situation is illustrated in Figure 2-5, which indicates how the various propagation mechanisms affect the measurable parameters of a signal on a communications link. The parameters that can be observed or measured on a typical link are amplitude, phase, polarization, frequency, bandwidth, and angle of arrival. Each of the propagation mechanisms, if present in the path, will affect one or more of the signal parameters, as shown on the figure. Since all of the signal parameters, except for frequency, can be affected by several mechanisms, it is usually not possible to determine the propagation conditions from an observation of the parameters alone, for example, if a reduction in signal amplitude is observed, it could be caused by absorption, scattering, refraction, diffraction, multipath, scintillation, fading, or a combination of the above.

Propagation effects on communications links are usually defined in terms of variations in the signal parameters, hence one or several mechanisms could be

*The term dispersion is also used to denote the differential delay experienced across the bandwidth of a radiowave propagating through a medium of free electrons, such as the ionosphere or a plasma.

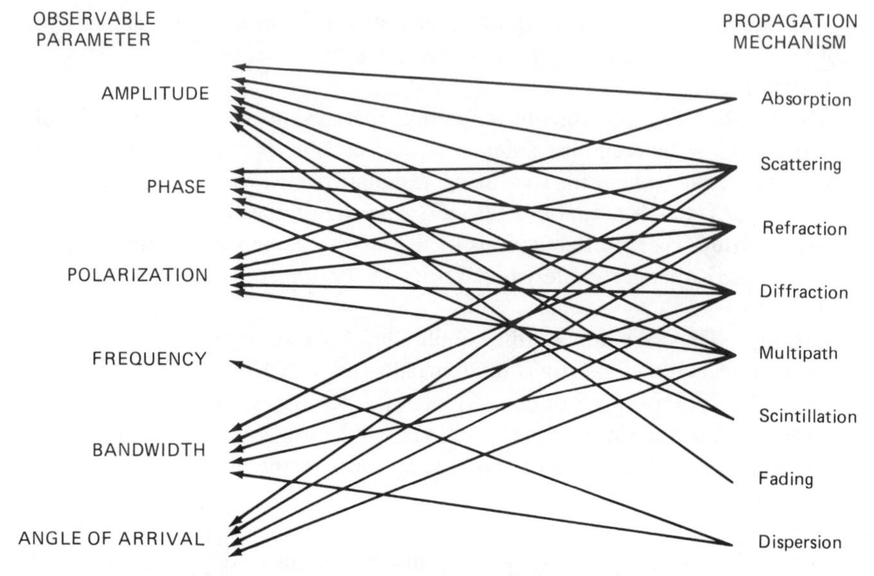

Figure 2-5. Radiowave propagation mechanisms and their impact on the parameters of a communications signal.

present on the link. A reduction of signal amplitude caused by rain in the path, for example, is the result of both absorption and scattering. As we proceed through our discussion of propagation factors in this and succeeding chapters, it will be helpful to recall the distinction between the propagation *effect* on a signal parameter and the propagation *mechanisms* which produce the variation in the parameter.

2.6. MAJOR RADIOWAVE PROPAGATION FACTORS IN SPACE COMMUNICATIONS

We have seen that frequency plays a major role in the determination of the propagation characteristics of radiowaves used for space telecommunications, and that the ionosphere is a critical element in the evaluation of propagation effects. It is useful, therefore, to divide the introductory discussion of radiowave propagation factors found in space communications into two regions of the frequency spectrum: a higher frequency region where the effects are primarily produced in the troposphere, i.e., where the ionosphere is essentially transparent to the radiowave, and a lower frequency region where the effects are determined by the ionosphere.

The breakpoint between the two regions is not at a specific frequency, but generally will occur at around 3 GHz (10 cm wavelength). There is some overlap for certain propagation conditions, and these will be discussed when appropriate.

As we have seen from the discussion of communications satellites and frequency allocations in Chapter 1, most current earth–space communications satellites operate in the frequency bands above 3 GHz, and the trend is to higher frequencies in the future. Our study of radiowave propagation effects on space communications will therefore be most concerned with those factors produced in the troposphere, or non-ionized portions of the Earth's atmosphere. There are several ionospheric propagation factors which are important for space communications, however, particularly in the mobile satellite bands located between 120 and 1660 MHz, and they are also included in the discussion.

In the following two sections (2.6.1 and 2.6.2) the major propagation factors in space communications found in the two frequency regions discussed above will be introduced. Later chapters will discuss the factors in more detail and will emphasize their impact on communications link design and performance.

2.6.1. Propagation Factors Above About 3 GHz

The major propagation factors which can hinder space communications in the frequency bands above about 3 GHz are introduced and briefly described in this section. The appropriate chapter where the subject is described in further detail is indicated after each description.

Gaseous Attenuation. A reduction in signal amplitude caused by the gaseous constituents of the Earth's atmosphere which are present in the transmission path. Gaseous attenuation is an absorption process, and the primary constituents of importance at space communcations frequencies are oxygen and water vapor. Gaseous attenuation increases with increasing frequency, and is dependent on temperature, pressure, and humidity. (See Chapter 3.)

Hydrometeor Attenuation. A reduction in signal amplitude caused by hydrometeors (rain, clouds, fog, snow, ice) in the transmission path. Hydrometeors are the products formed by the condensation of atmospheric water vapor. Hydrometeor attenuation experienced by a radiowave involves both absorption and scattering processes. *Rain attenuation* can produce major impairments in space communications, particularly in the frequency bands above 10 GHz. *Cloud and fog attenuation* is much less severe than rain attenuation; however, it must be considered in link calculations, particularly for frequencies above 15 GHz. *Dry snow and ice particle attenuation* is usually so low that it is unobservable on space communications links operating below 30 GHz. (See Chapters 4 and 5.)

Depolarization. A change in the polarization characteristics of a radiowave caused by (a) hydrometeors, primarily rain or ice particles; and (b) multipath propagation. A depolarized radiowave will have its polarization state altered such that power is transferred from the desired polarization state to an undesired orthogonally polarized state, resulting in interference or crosstalk between the two orthogonally polarized channels. Rain and ice depolarization can be a problem in the frequency bands above about 12 GHz, particularly for "frequency reuse" communications links which employ dual independent orthogonal polarized channels in the same frequency band to increase channel capacity. Multipath depolarization is generally limited to very low elevation angle space communcations, and will be dependent on the polarization characteristics of the receiving antenna. (See Chapter 6.)

Radio Noise. The presence of undesired signals or power in the frequency band of a communications link, caused by natural or man-made sources. Radio noise can degrade the noise characteristics of receiver systems and affect antenna design or system performance. The primary natural noise sources for frequencies above about 1 GHz are: atmospheric gases (oxygen and water vapor), rain, clouds, and surface emmissions. Man-made sources include: other space or terrestrial communications links, electrical equipment, and radar systems. Extraterrestial cosmic noise must only be considered for frequencies below about 1 GHz. (See Chapter 7.)

Angle of Arrival Variations. A change in the direction of propagation of a radiowave caused by refractive index changes in the transmission path. Angle of arrival variations are a refraction process, and generally are only observable with large aperture antennas (10 meters or more), and at frequencies well above 10 GHz. The angle of arrival change results in an apparent shift in the location of satellite position, and can be compensated for by a repointing of the antenna. (See Chapter 8.)

Bandwidth Coherence. An upper limit on the information bandwidth or channel capacity that can be supported by a radiowave, caused by the dispersive properties of the atmosphere, or by multipath propagation. The *coherence bandwidth* for typical space communcation frequencies is one or more gigahertz, and is not expected to be a severe problem, except for links which must propagate through a plasma. (See Chapter 8.)

Antenna Gain Degradation. An apparent reduction in the gain of a receiving antenna caused by amplitude and phase decorrelation across the aperture. This effect can be produced by intense rain; however, it is only observable with very large aperture antennas at frequencies above about 30 GHz and for very long path lengths through the rain, i.e., low elevation angles. (See Chapter 8.)

2.6.2. Propagation Factors Below About 3 GHz

The major propagation factors which can affect space communications at frequencies above the ionospheric penetration frequency and up to about 3 GHz are introduced in this section. The factors are presented in approximate order of decreasing importance to space communications system design and performance.

Ionospheric Scintillation. Rapid fluctuations of the amplitude and phase of a radiowave, caused by electron density irregularities in the ionosphere. Scintillation effects have been observed on links from 30 MHz to 7 GHz, with the bulk of observations of amplitude scintillation in the VHF (30–300 MHz) band [2.2]. The scintillations can be very severe and can determine the practical limitation for reliable communications under certain atmospheric conditions. Ionospheric scintillations are most severe for transmission through equatorial, auroral, and polar regions; and during sunrise and sunset periods of the day.

Polarization Rotation. A rotation of the polarization sense of a radiowave, caused by the interaction of a radiowave with electrons in the ionosphere, in the presence of the Earth's magnetic field. This condition, referred to as the *Faraday Effect*, can seriously affect VHF space communications systems which use linear polarization. A rotation of the plane of polarization occurs because the two rotating components of the wave progress through the ionosphere with different velocities of propagation. Faraday rotations of 30 revolutions (10,800 degrees) can occur at 100 MHz, with the effect decreasing with increasing frequency by the reciprical of the frequency squared.

Group Delay (or Propagation Delay). A reduction in the propagation velocity of a radiowave, caused by the presence of free electrons in the propagation path. The group velocity of a the radiowave is retarded (slowed down), thereby increasing the travel time over that expected for a free space path. This effect can be extremely critical for radionavigation or satellite ranging links which require an accurate knowledge of range and propagation time for successful performance. Group delay will be about 25 microseconds at 100 MHz for an earth–space path at a 30 degree elevation angle, and is approximately proportional to the reciprical of the frequency squared.

Multipath Fading and Scintillation. Variations in the amplitude and phase of a radiowave, caused by terrain and surface roughness conditions. This problem is important in terrestrial communications and must also be considered for earth–space transmissions at low elevation angles, and for VHF mobile satellite links.

Tropospheric Refraction and Fading. Changes in the angle of arrival or the amplitude of a radiowave, caused by tropospheric refractive index variations. The index of refraction of the troposphere at radio frequencies is a function of temperature, pressure, and water vapor content. Tropospheric refractive bending and amplitude fading can occur at frequencies above and below 3 GHz, but the problem is most pronounced at low elevation angles, i.e., 5–10 degrees.

Radio Noise. See description in prior section.

The factors discussed above are treated in more detail in Chapter 8. Several of the factors are also discussed in various sections which deal with propagation considerations for specific earth–space applications.

REFERENCES

2.1. IEEE Standard Definitions of Terms for Radio Wave Propagation, IEEE Std. 211-1977, New York, August 19, 1977.

2.2. CCIR, Report 263-4, "Ionospheric Effects Upon Earth-space Propagation," in Volume VI, *Propagation in Ionized Media*, Recommendations and Reports of the CCIR, 1978, pp. 71–89, International Telecommunications Union, Geneva, 1978.

CHAPTER 3
ATTENUATION BY ATMOSPHERIC GASES

A radiowave propagating through the Earth's atmosphere will experience a reduction in signal level due to the gaseous components present in the transmission path. Signal degradation can be minor or severe, depending on frequency, temperature, pressure, and water vapor concentration. In this chapter the effects of atmospheric gases, primarily oxygen and water vapor at space communications frequencies, are discussed, and methods for calculating the expected attenuation for a radiowave link are presented. Examples are given for atmospheric attenuation at several of the frequencies of interest for space communications systems. Atmospheric gases also affect radio communications by adding atmospheric noise (i.e., radio noise) to the link. This problem is described fully in Chapter 7.

3.1. OXYGEN AND WATER VAPOR ATTENUATION

There are many gaseous constituents in the Earth's atmosphere which can interact with a radiowave link. The principal components of the dry atmosphere, and their approximate percentage by volume, are: oxygen, 21%; nitrogen, 78%; argon, 0.9%; and carbon dioxide, 0.1%—all well mixed to a height of about 80 km [3.1]. Water vapor is the principal variable component of the atmosphere, and at sea level and 100% relative humidity it constitutes about 1.7% by volume of the U.S. Standard Atmosphere, 1976 [3.2].

The principal interaction mechanism involving the gaseous constituents and a radiowave is molecular absorption, which results in a reduction in signal amplitude (attenuation) of the radiowave. The absorption of the radiowave results from a quantum level change in the rotational energy of the molecule, and occurs at a specific resonant frequency or narrow band of frequencies. The resonant frequency of interaction depends on the energy levels of the initial and final rotational energy states of the molecule.

Only oxygen and water vapor have observable resonance frequencies in the bands of interest for space communications. Oxygen has a series of very close absorption lines near 60 GHz and an isolated absorption line at 118.74 GHz. Water vapor has lines at 22.3 GHz, 183.3 GHz, and 323.8 GHz. Oxygen absorption involves magnetic dipole changes, while water vapor absorption consists of electric dipole transitions between rotational states.

The attenuation produced by oxygen and water vapor in the atmosphere is described by the specific attenuation for each component, expressed in dB/km. (This expression is sometimes referred to as an attenuation coefficient). The specific attenuation is found by summing the contributions of each quantum level transition for the molecule. The calculation is a complex evaluation for each value of temperature, pressure, and water vapor concentration [3.3, 3.4]. Approximate techniques are available with sufficient accuracy, however, for the determination of specific attenuation due to atmospheric oxygen and water vapor for most practical communications applications.

One technique employs an approximation based on the absorption line shape profiles of Van Vleck and Weisskopf [3.5]. The specific attenuation for oxygen, γ_o, and for water vapor, γ_w, at 20°C surface temperature, are determined from the following expressions:

For $f < 57$ Ghz:

$$\gamma_o(\text{dB/km}) = \left[\frac{6.6}{f^2 + 0.33} + \frac{9}{(f - 57)^2 + 1.96}\right] f^2\ 10^{-3} \tag{3-1a}$$

For $57 \leq f \leq 63$ GHz:

$$\gamma_o(\text{dB/km}) = 14.9 \tag{3-1b}$$

For $63 < f \leq 350$ GHz:

$$\gamma_o(\text{dB/km}) = \left[\frac{4.13}{(f - 63)^2 + 1.1} + \frac{0.19}{(f - 118.7)^2 + 2}\right] f^2\ 10^{-3} \tag{3-1c}$$

For $f \leq 350$ GHz:

$$\gamma_w(\text{dB/km}) = \left[0.067 + \frac{2.4}{(f - 22.3)^2 + 6.6} + \frac{7.33}{(f - 183.5)^2 + 5} + \frac{4.4}{(f - 323.8)^2 + 10}\right] f^2 \rho\ 10^{-4} \tag{3-2}$$

where f is the frequency in GHz and ρ is the water vapor concentration in g/m^3.

Figure 3-1 presents the values of specific attenuation at frequencies from 3 to 350 GHz obtained from the above expressions for a water vapor concentra-

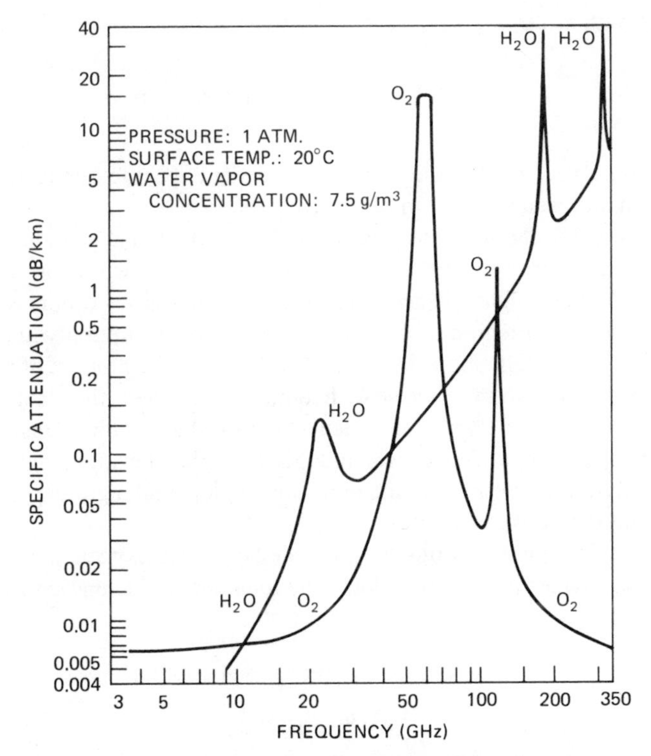

Figure 3.1. Specific attenuation for atmospheric oxygen (O_2) and water vapor (H_2O) as a function of frequency.

tion of 7.5 g/m^3, which corresponds to a relative humidity of 42% at 20°C. The flat peak at the 60 GHz O_2 line, Equation (3-1b), represents the average value for a large number of lines which occur in the 57–63 GHz region.

The above results are developed for a surface temperature of 20°C, and the specific attenuation will increase slightly as the temperature decreases. The effect of temperature may be taken into account by adding a factor of

$$\Delta\gamma(\text{dB/km}) = 0.01(20 - T_0) \tag{3-3}$$

to the equations for γ_o and γ_w, where T_0 is the surface temperature in degrees Celsius. Note that as the temperature decreases, the attenuation coefficient will increase by about 1% per degree Celsius, and vice versa.

3.2. TOTAL SLANT PATH ATMOSPHERIC ATTENUATION

The total atmospheric attenuation at a given frequency, A_a, for a path length of r_0 kilometers, is determined by integration of the specific attenuation values along the path r, i.e.,

$$A_a = \int_0^{r_0} [\gamma_o(r) + \gamma_w(r)]\, dr, \quad \text{dB} \tag{3-4}$$

where $\gamma_o(r)$ and $\gamma_w(r)$ are the position dependent specific attenuation profiles for oxygen and water vapor, in dB/km.

The total atmospheric attenuation for a slant path at an elevation angle θ can be determined from the specific attenuation values in two steps. First an atmospheric "scale height" is assumed for oxygen and water vapor, which in the effective interaction region in the atmosphere where absorption will occur. The zenith ($\theta = 90°$) attenuation, in dB, is then found as the product of the specific attenuation, in dB/km, and the scale height, in km. Then, the slant path attenuation at an elevation angle θ is calculated from the zenith attenuation by an appropriate path length factor, which depends on the elevation angle. The general procedure for determining the elevation angle path length dependence factor for slant path links is given in Appendix A.

For example, consider a link with a scale height for oxygen assumed to be 8 km, and for water vapor to be 2 km. The total zenith attenuation for this case is

$$\begin{aligned} A_T(90°) &= A_o(90°) + A_w(90°) \\ &= 8\gamma_o + 2\gamma_w \end{aligned} \tag{3-5}$$

where γ_o and γ_w are determined from Equations (3-1) and (3-2).

Then, from the results of Appendix A, the attenuation for an elevation angle θ will be: For $\theta > 10°$,

$$\begin{aligned} A_T(\theta) &= \frac{A_o(90°)}{\sin\theta} + \frac{A_w(90°)}{\sin\theta} \\ &= \frac{8\gamma_o}{\sin\theta} + \frac{2\gamma_w}{\sin\theta} \end{aligned}$$

and for $\theta < 10°$,

$$\begin{aligned} A_T(\theta) &= \frac{2A_o(90°)}{\sqrt{\sin^2\theta + \frac{2(8)}{R}} + \sin\theta} + \frac{2A_w(90°)}{\sqrt{\sin^2\theta + \frac{2(2)}{R}} + \sin\theta} \\ &= \frac{16\gamma_o}{\sqrt{\sin^2\theta + (16/R)} + \sin\theta} + \frac{4\gamma_w}{\sqrt{\sin^2\theta + (4/R)} + \sin\theta} \end{aligned}$$

where R is the effective radius of the earth, usually assumed to be 8500 km.

3.2.1. Multiple Regression Analysis Procedure

A more useful and applicable procedure for the determination of total slant path attenuation for space communications links has been developed from direct measurements of oxygen and water vapor profiles [3.6]. This procedure provides the total attenuation at any location and elevation angle, based on local surface temperature and water vapor concentration. A multiple regression analysis was performed on a selected global sample of 220 radiosonde profiles representing all seasons and geographic locations. The regression analysis provides a *combined* specific attenuation, γ_a, representing the contributions of both oxygen and water vapor absorption, which is statistically related to the surface water vapor concentration and surface temperature through a set of empirical frequency dependent coefficients. The total zenith atmospheric attenuation, A_a, is also determined by the regression analysis through a second set of empirical coefficients which are dependent on operating frequency also.

The specific attenuation γ_a is computed from the relation,

$$\gamma_a = a(f) + b(f)\rho_0 - c(f)T_0, \quad \text{dB/km} \tag{3-6}$$

and the total zenith ($\theta = 90°$) atmospheric attenuation A_a from

$$A_a(90°) = \alpha(f) + \beta(f)\rho_0 - \xi(f)T_0, \quad \text{dB} \tag{3-7}$$

where ρ_0 is the mean local surface water vapor concentration in g/m^3, T_0 is the mean local surface temperature in °C, and $a(f)$, $b(f)$, $c(f)$, $\alpha(f)$, $\beta(f)$, and $\xi(f)$ are frequency dependent empirical coefficients developed from the multiple regression analysis. Table 3-1 presents the regression coefficients for specific attenuation, Equation (3-6), and Table 3-2 presents the coefficients for total zenith atmospheric attenuation, Equation (3-7), at selected frequencies from 1 to 350 GHz. To determine the coefficients for other frequencies not in the tables, a power law interpolation is required. The procedure for coefficient interpolation is described in Appendix B.

For slant path calculations, the scale height H_a is found from the regression coefficients as,

$$H_a = \frac{A_a(90°)}{\gamma_a} = \frac{\alpha(f) + \beta(f)\rho_0 - \xi(f)T_0}{a(f) + b(f)\rho_0 - c(f)T_0} \tag{3-8}$$

Figure 3-2 shows a plot of total one-way zenith atmospheric attenuation calculated from the regression coefficient method, for a ρ_0 of 7.5 g/m^3 and a T_0 of

Table 3-1. Coefficients for the Calculation of Specific Attenuation Due to Gaseous Absorption.

$$\gamma_a = a(f) + b(f)\,\rho_0 - c(f)\,T_0$$

Frequency f(GHz)	Coefficients		
	$a(f)$	$b(f)$	$c(f)$
1	0.00588	0.0000178	0.0000517
4	0.00802	0.000141	0.0000850
6	0.00824	0.000300	0.0000895
12	0.00898	0.00137	0.000108
15	0.00953	0.00269	0.000125
16	0.00976	0.00345	0.000133
20	0.0125	0.0125	0.000101
22	0.0181	0.0221	0.000129
24	0.0162	0.0203	0.0000563
30	0.0179	0.0100	0.000280
35	0.0264	0.0101	0.000369
41	0.0499	0.0121	0.000620
45	0.0892	0.0140	0.00102
50	0.267	0.0171	0.00251
55	3.93	0.0220	0.0158
70	0.449	0.0319	0.00443
80	0.160	0.0391	0.00130
90	0.113	0.0495	0.000744
94	0.106	0.0540	0.000641
110	0.116	0.0749	0.000644
115	0.206	0.0826	0.00185
120	0.985	0.0931	0.0115
140	0.123	0.129	0.000372
160	0.153	0.206	0.000784
180	1.13	1.79	−0.00237
200	0.226	0.366	0.00167
220	0.227	0.316	0.000174
240	0.258	0.356	−0.000119
280	0.336	0.497	−0.0000664
300	0.379	0.629	0.000808
310	0.397	0.812	0.00286
320	0.732	2.36	0.00467
330	0.488	1.61	0.00945
340	0.475	1.06	0.00519
350	0.528	1.23	0.00722

NOTE: See Appendix B for interpolation procedure for this table.

Table 3-2. Coefficients for the Calculation of Total Zenith (θ = 90°) Atmospheric Attenuation.

$$A_a(90°) = \alpha(f) + \beta(f)\rho_0 - \xi(f)T_0$$

Frequency f(GHz)	Coefficients		
	$\alpha(f)$	$\beta(f)$	$\epsilon(f)$
1	0.0334	0.00000276	0.000112
4	0.0397	0.000276	0.000176
6	0.0404	0.000651	0.000196
12	0.0436	0.00318	0.000315
15	0.0461	0.00634	0.000455
16	0.0472	0.00821	0.000536
20	0.0560	0.0346	0.00155
22	0.0760	0.0783	0.00310
24	0.0691	0.0591	0.00250
30	0.0850	0.0237	0.00133
35	0.123	0.0237	0.00149
41	0.237	0.0284	0.00211
45	0.426	0.0328	0.00299
50	1.27	0.0392	0.00572
55	24.5	0.0490	−0.00121
70	2.14	0.0732	0.0104
80	0.705	0.0959	0.00586
90	0.458	0.122	0.00574
94	0.417	0.133	0.00594
110	0.431	0.185	0.00785
115	0.893	0.203	0.0113
120	5.35	0.221	0.0363
140	0.368	0.319	0.0119
160	0.414	0.506	0.0191
180	2.81	5.04	0.192
200	0.562	0.897	0.0339
220	0.543	0.777	0.0276
240	0.601	0.879	0.0307
280	0.760	1.22	0.0428
300	0.853	1.54	0.0551
310	0.905	1.97	0.0735
320	1.66	6.13	0.238
330	1.13	3.94	0.155
340	1.07	2.56	0.0969
350	1.20	2.96	0.114

NOTE: See Appendix B for interpolation procedure for this table.

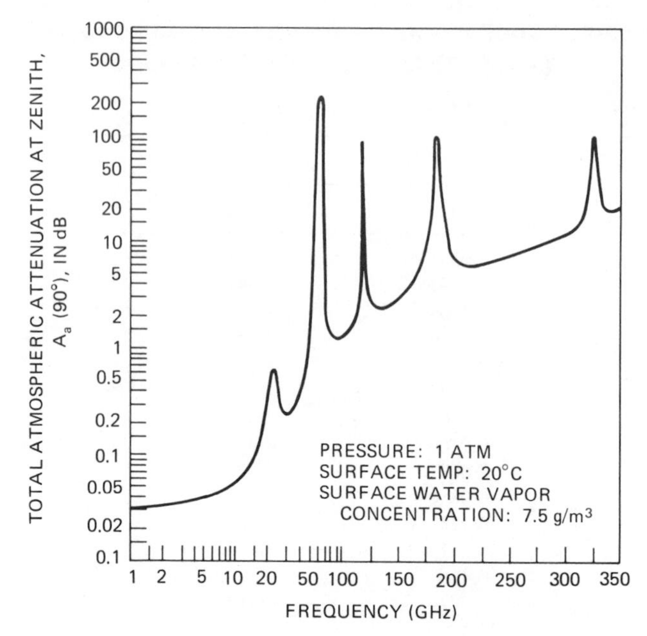

Figure 3.2. Total atmospheric gaseous attenuation at zenith, calculated from regression coefficient method.

20°C. This plot vividly points out the atmospheric windows where earth–space communications are practical, and the absorption regions where communications are not practical. The attenuation is less than 1 dB up to about 50 GHz, where it rapidly increases to over 100 dB at the oxygen absorption band around 60 GHz. A second window occurs around 90 GHz and a third between 120 and 150 GHz.

The accuracy of the multiple regression method is heavily dependent on the mean surface values of ρ_0 and T_0 used in the calculations, as well as the deviation inherent in the regression analysis. The variation in the total attenuation calculation to be expected at a single location can be estimated from,

$$\sigma_{A_a}^2 = \beta^2(f)\,\sigma_{\rho 0}^2 + \xi^2(f)\,\sigma_{T_0}^2 + \sigma_m^2 \qquad (3\text{-}9)$$

where σ_{A_a} is the standard deviation of the total attenuation A_a; $\sigma_{\rho 0}$ and σ_{T_0} are the standard deviations of water vapor concentration and temperature at that location, respectively; and σ_m is the standard deviation about the estimate from the regression analysis.

For example, the expected total zenith attenuation estimate and standard deviations for a link with a ρ_0 of 11.1 g/m^3 and a T_0 of 15°C, at 20, 30 and 42 GHz are found as follows [3.6].

	20 GHz	30 GHz	42 GHz
$A_a(90°)$ →	0.17 dB	0.35 dB	0.5 dB
σ_{A_a} →	0.07 dB	0.2 dB	0.16 dB
σ_m →	0.028 dB	0.04 dB	0.024 dB

The total uncertainty of the estimates would therefore be 0.17 ± 0.07 dB, 0.35 ± 0.2 dB, and 0.5 ± 0.16 dB, respectively, if average values of ρ_0 and T_0 are used. If local conditions are directly known, however, the uncertainty is reduced to the value of σ_m only, i.e., 0.17 ± 0.028 dB, 0.35 ± 0.04 dB, and 0.5 ± 0.024 dB, which corresponds to about a 12–15% overall uncertainty in the estimate.

A comparison of the multiple regression technique with radiative transfer calculations for the specific attenuation coefficients in the first four window regions shows an average difference in the magnitudes of 5% in dB [3.7]. That same reference also reports comparisons with measured values of zenith attenuation which were within 15% for frequencies below 70 GHz.

Tables 3-3a, 3-3b, and 3-3c list the total one-way gaseous attenuation as a function of elevation angle calculated by the multiple regression method for several of the space communications bands below 100 GHz. Three types of atmospheric conditions are presented: dry atmosphere, moderate, and hot-humid, with the values for T_0, ρ_0, and relative humidity as stated on the tables.

Table 3-3a. Total Gaseous Attenuation for Satellite Paths Through the Atmosphere–Dry Atmosphere.

T_0 = 20°C
ρ_0 = 0.001 g/m^3
Rel. Humidity = 10%

Frequency (GHz)	Elevation Angle 0°	5°	10°	30°	45°	90°
1	10.30	2.19	1.15	0.40	0.28	0.03
2	11.18	2.45	1.29	0.45	0.32	0.03
4	11.29	2.28	1.19	0.41	0.29	0.04
6	11.31	2.27	1.19	0.41	0.29	0.04
12	11.37	2.25	1.17	0.41	0.29	0.04
15	11.07	2.15	1.12	0.39	0.28	0.04
20	5.04	0.67	0.34	0.12	0.08	0.03
30	16.60	3.07	1.60	0.55	0.39	0.06
41	57.89	11.18	5.83	2.02	1.43	0.19
50	347.87	67.97	35.47	12.32	8.71	1.16
94	69.58	10.69	5.50	1.91	1.35	0.30

Table 3-3b. Total Gaseous Attenuation for Satellite Paths Through the Atmosphere—Moderate.

$T_0 = 20°C$
$\rho_0 = 7.5\ g/m^3$
Rel. Humidity = 42%

Frequency (GHz)	Elevation Angle					
	0°	5°	10°	30°	45°	90°
1	10.17	2.14	1.12	0.39	0.28	0.03
2	10.88	2.31	1.22	0.42	0.30	0.03
4	11.36	2.19	1.14	0.40	0.28	0.04
6	11.76	2.18	1.13	0.39	0.28	0.04
12	15.08	2.44	1.26	0.44	0.31	0.06
15	19.43	2.95	1.51	0.53	0.37	0.08
20	61.28	8.73	4.47	1.55	1.10	0.28
30	50.64	7.18	3.68	1.28	0.90	0.24
41	94.81	14.53	7.47	2.59	1.83	0.41
50	387.40	67.74	35.07	12.18	8.61	1.45
94	272.45	37.92	19.41	6.74	4.77	1.30

Table 3-3c. Total Gaseous Attenuation for Satellite Paths Through the Atmosphere—Hot-Humid.

$T_0 = 30°C$
$\rho_0 = 18\ g/m^3$
Rel. Humidity = 60%

Frequency (GHz)	Elevation Angle					
	0°	5°	10°	30°	45°	90°
1	9.98	2.13	1.12	0.39	0.28	0.03
2	10.49	2.22	1.16	0.40	0.29	0.03
4	11.39	2.14	1.12	0.39	0.27	0.04
6	12.39	2.17	1.12	0.39	0.28	0.05
12	20.66	3.08	1.58	0.55	0.39	0.09
15	31.43	4.46	2.28	0.79	0.56	0.15
20	135.38	19.17	9.82	3.41	2.41	0.63
30	97.03	13.22	6.76	2.35	1.66	0.47
41	148.07	21.17	10.84	3.77	2.66	0.68
50	447.00	72.77	37.52	13.03	9.21	1.80
94	541.32	73.72	37.70	13.09	9.26	2.63

The 50 GHz value is included as an example of the extensive attenuation observed in an absorption band. Atmospheric attenuation, in general, increases with frequency, with the notable exception of the 30 and 20 GHz bands. For those conditions where some water vapor contribution is present, the 20 GHz band will have a higher attenuation than the 30 GHz band because of the closer proximity of the 22 GHz absorption line. This interesting situation is often overlooked in 30/20 GHz system designs, where it is assumed that the higher frequency will have the higher atmospheric attentuation.

The dry atmosphere case consists primarily of oxygen absorption. Note the low values up through 30 GHz because of the lack of the first water vapor line contribution.

The moderate case, which is typical for most temperate regions in the United States, shows relatively low (about 1 dB) values for satellite link elevation angles of 30° or higher. Only at the low elevation angles will atmospheric absorption be a critical design factor.

The hot-humid atmosphere represents an upper limit for atmospheric absorption contributions for the United States, and the results are only slightly higher than the moderate atmosphere. When compared to hydrometeor attenuation and other effects, atmospheric absorption is generally minor by comparison, except for very low elevation angle applications in humid environments.

3.3. SUMMARY OF ATMOSPHERIC ATTENUATION PROCEDURE

The multiple regression analysis, as we have seen, offers the most direct and practical procedure for a reasonable estimate of total atmospheric attenuation due to gaseous absorption in an earth-space link. This section summarizes the steps in the procedure for the calculation of total atmospheric attenuation at an elevation angle θ and operating frequency f_0.

Required Input Information:

Operating Frequency: f_0, in GHz
Elevation Angle: θ, in Degrees
Surface Water Vapor Concentration: ρ_0, in g/m^3
Surface Temperature: T_0, in degrees Celsius

STEP 1: Calculate the specific attenuation γ_a using Equation (3-6), at the desired frequency f_o, with the coefficients of Table 3-1 (use interpolation procedure of Appendix B if required). That is,

$$\gamma_a(f_0) = a(f_0) + b(f_0)\rho_0 - c(f_0)T_0$$

STEP 2: Determine the total zenith attenuation $A_a(90°)$ from Equation (3-7), with the coefficients of Table 3-2 (use interpolation procedure of Appendix B if required). That is,

$$A_a(90°)|_{f0} = \alpha(f_0) + \beta(f_0)\rho_0 - \xi(f_0)T_0$$

STEP 3: For slant-path calculations, determine the scale height H_a from Equation (3-8), i.e.,

$$H_a = \frac{A_a(90°)|_{f0}}{\gamma_a(f_0)}$$

STEP 4: The total one-way atmospheric attenuation at the slant-path elevation angle θ is then found: For $\theta \geq 10°$,

$$A_a(\theta) = \frac{H_a A_a\,(90°)}{\sin\theta}$$

or, for $\theta < 10°$

$$A_a(\theta) = \frac{2H_a A_a\,(90°)}{\sqrt{\sin^2\theta + \frac{2H_a}{8500}} + \sin\theta}$$

Note also that

$$A_a(0) = \sqrt{2H_a R}\,A_a\,(90°)$$
$$= 130.38\,\sqrt{H_a}\,A_a\,(90°)$$

This procedure can be expected to provide estimates to within 10 to 15% for operating frequencies below about 70 GHz. For higher frequencies the more detailed techniques employing radiative transfer calculations are recommended.

REFERENCES

3.1. Battan, L. J., *Fundamentals of Meteorology*, Prentice-Hall, Englewood Cliffs, N.J., 1979.
3.2. "U.S. Standard Atmosphere, 1976," NDAA–S/T 76-1562, U.S. Government Printing Office, Washington, D.C., October 1976.
3.3. Van Vleck, J. H. and Weisskopf, V. F., "On the shape of Collision Broadened Lines," *Rev. Mod. Phys.*, Vol. 17, pp. 227–236, 1945.
3.4. Liebe, H. J., "Modeling attenuation and phase of radio waves in air at frequencies below 1000 GHz," *Radio Science*, Vol. 16, No. 6, pp. 1183–1199, Nov.–Dec., 1981.

3.5. CCIR, Report 719-1, ''Attenuation By Atmospheric Gases,'' in Volume V, *Propagation in Non-ionized Media*, Recommendations and Reports of the CCIR, 1982, pp. 138–150, International Telecommunications Union, Geneva, 1982.
3.6. Crane, R. K., ''An Algorithm to Retrieve Water Vapor Information from Satellite Measurements'' NEPRF Tech. Rept. 7-76 (ERT), Final Report, Project No. 1423, Environmental Research and Technology, Inc., November 1976.
3.7. Smith, Ernest K., ''Centimeter and millimeter wave attenuation and brightness temperature due to atmospheric oxygen and water vapor,'' *Radio Science*, Vol. 17, No. 6, pp. 1455–1464, Nov.-Dec., 1982.

CHAPTER 4
HYDROMETEOR ATTENUATION ON SATELLITE PATHS

The effects of precipitation on the transmission path are of major concern in space communications, particularly for those systems operating at frequencies above 10 GHz. Precipitation can have many forms in the atmosphere. Hydrometeor is the general term referring to the products of condensed water vapor in the atmosphere, observed as rain, clouds, fog, hail, ice, or snow. The presence of hydrometeors in the radiowave path, particularly rain, can produce major impairments to space communications. Rain drops absorb and scatter radiowave energy, resulting in signal attenuation (a reduction in the transmitted signal amplitude), which can degrade the reliability and performance of the communications link.

This chapter presents a review of rain, cloud, and fog attenuation on satellite links. (The other hydrometeors—hail, ice, and snow, play a very minor role in producing attenuation on radiowave paths.) The classical derivation for rain attenuation is presented, along with recent direct measurements on satellite paths. The latest developments in rain attenuation prediction and modeling are discussed in Chapter 5. Other effects caused by hydrometeors, such as depolarization, radio noise, scintillations, and bandwidth reduction, are covered in later chapters.

4.1. CLASSICAL DEVELOPMENT FOR RAIN ATTENUATION

The classical development for the determination of radiowave attenuation due to rain began with studies by early investigators in the years immediately following World War II [4.1, 4.2, 4.3]. Later, in 1965, Medhurst [4.4] compared the theory with extensive sets of terrestrial measurements and produced an empirical procedure for calculating rain attenuation at centimeter wavelengths. The development has proceeded from these early studies to the present, with further enhancements and improvements, most recently with the addition of slant paths to allow for space communications links.

The classical development for rain attenuation is based on three assumptions describing the nature of radiowave propagation and precipitation.

1. The intensity of the wave *decays exponentially* as it propagates through the volume of rain.
2. The rain drops are assumed to be *spherical* water drops, which both scatter and absorb energy from the incident radiowave.
3. The contributions of each drop are *additive* and are *independent* of the other drops. This implies a "single scattering" of energy; however, as will be discussed later, the empirical results of the classical development do allow for some "multiple scattering" effects.

The determination of rain attenuation for a radiowave path proceeds from application of these assumptions.

The attenuation of a radiowave propagating in a volume of rain of extent L in the direction of wave propagation can be expressed as

$$A = \int_0^L \alpha \, dx \tag{4-1}$$

where α is the specific attenuation of the rain volume, expressed in dB/km, and the integration is taken along the extent of the propagation path, from $x = 0$ to $x = L$.

Consider a plane wave of transmitted power P_t incident on a volume of uniformly distributed spherical water drops, all of radius r, extending over a length L, as shown in Figure 4-1. The received power P_r will be

$$P_r = P_t \, e^{-kL} \tag{4-2}$$

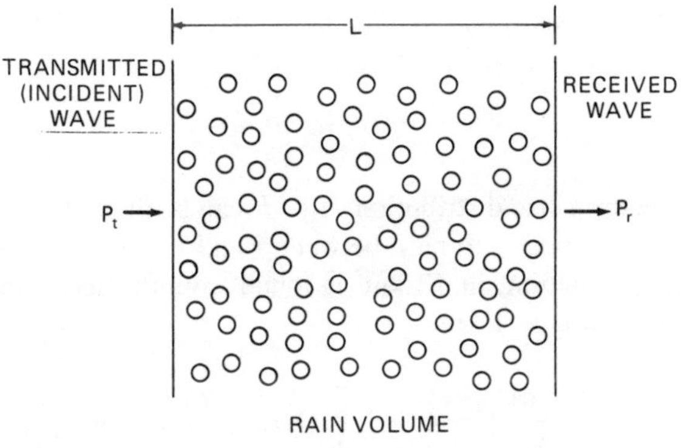

Figure 4-1. Radiowave incident on a volume of spherical, uniformly distributed water drops.

where k is the attenuation coefficient for the rain volume, expressed in units of reciprocal length.

The attenuation of the wave, usually expressed as a positive decibel (dB) value, is given by

$$A(\text{dB}) = 10 \log_{10} \frac{P_t}{P_r} \tag{4-3}$$

Converting the logarithm to the base e and employing Equation (4-2),

$$A(\text{dB}) = 4.343 kL \tag{4-4}$$

The attenuation coefficient k is expressed as

$$k = \rho Q_t \tag{4-5}$$

where ρ is the drop density, i.e., the number of drops per unit volume, and Q_t is the attenuation cross section of the drop, expressed in units of area. Q_t is the sum of a scattering cross section Q_s and an absorption cross section Q_a. The attenuation cross section is a function of the drop radius, r, wavelength of the radiowave, λ, and complex refractive index of the water drop, m. That is

$$Q_t = Q_s + Q_a = Q_t(r, \lambda, m) \tag{4-6}$$

The concept of cross section, developed in early radar studies, describes the physical profile that an object projects to a radiowave. It is defined as the ratio of the total power extracted from the wave (in watts) to the total incident power density (in watts per square meter); hence the unit of area, square meter.

The drops in a real rain are not all of uniform radius, and the attenuation coefficient must be determined by integrating over all of the drop sizes, i.e.,

$$k = \int Q_t(r, \lambda, m)\, n(r)\, dr \tag{4-7}$$

where $n(r)$ is the drop size distribution. $n(r)\, dr$ can be interpreted as the number of drops per unit volume with radii between r and $r + dr$.

The specific attenuation, in dB/km, is found from the above result for k and Equation (4-4), with $L = 1$ km,

$$\alpha \left(\frac{\text{dB}}{\text{km}}\right) = 4.343 \int Q_t(r, \lambda, m)\, n(r)\, dr \tag{4-8}$$

The above result demonstrates the dependence of rain attenuation on drop size, drop size distribution, rain rate, and attenuation cross section. The first three parameters are characteristics of the rain structure only. It is through the attenuation cross section that the frequency and temperature dependence of rain attenuation is determined. All of the parameters exhibit time and spatial variabilities which are not deterministic or directly predictable, hence most analyses of rain attenuation must rely on statistical analyses to quantatatively evaluate the impact of rain on communications systems.

The solution of Equation (4-8) requires Q_t and $n(r)$ as a function of the drop size. Q_t is found by employing the classical scattering theory of Mie for a plane wave radiating an absorbing sphere [4.5]. The resulting series expansion solution for Q_t is found [4.6] as follows:

$$Q_t = \frac{\lambda^2}{2\pi} \sum_{n=1}^{\infty} (2n + 1) \operatorname{Re} [a_n + b_n] \tag{4-9}$$

where a_n and b_n are the Mie scattering coefficients, which are complex functions of r, λ, and m. Re indicates "the real part of."

Equation (4-9), known as the *Mie scattering cross section*, is difficult to calculate; however, it can be simplified under the conditions where

$$2\pi r \ll \lambda \tag{4-10}$$

that is, where the size of the rain drop is much less than the wavelength of the radiowave. This condition, known as the *Rayleigh approximation*, is valid up to frequencies of 40–80 GHz. Under this condition, the attenuation cross section reduces to (see Appendix C)

$$Q_t = \frac{8\pi^2}{\lambda} r^3 \operatorname{Im} \left[\frac{m^2 - 1}{m^2 + 2}\right] \tag{4-11}$$

where Im indicates "the imaginary part of." Equation (4-11) is referred to as the *Rayleigh scattering cross section*.

Several investigators have studied the distributions of rain drop size as a function of rain rate and type of storm activity, and the drop size distributions were found to be well represented by an exponential of the form

$$n(r) = N_0 \, e^{-\Lambda r} = N_0 \, e^{-[cR^{-d}]r} \tag{4-12}$$

where R is the rain rate, in mm/h, and r is the drop radius, in mm. N_o, Λ, c, and d are empirical constants determined from the measured distributions.

Table 4-1. Drop Size Distributions Used for Rain Attenuation Calculations.

Drop Size Distribution	Empirical Constants: $N_0, \left(\frac{1}{\text{m}^2}\frac{1}{\text{mm}}\right)$	$\Lambda, \left(\frac{1}{\text{mm}}\right)$
Marshall–Palmer	8×10^3	$8.2R^{-.21}$
Joss:		
drizzle	30×10^3	$11.4R^{-.21}$
widespread rain	7×10^3	$8.2R^{-.21}$
thunderstorm	1.4×10^3	$6R^{-.21}$

NOTE: R = rain rate in mm/n.

The three most often referenced distributions are those of Laws and Parsons [4.7], Marshall and Palmer [4.8], and Joss and Waldvogel [4.9]. The Laws and Parsons distribution was measured directly and presented in tabular form for specific rain rate values from 0.25 to 150 mm/h. The Marshall–Palmer and Joss distributions were obtained from radar measurements and the empirical constants of Equation (4-12) for those distributions are listed in Table 4-1. The Joss distributions are classified into three rain types, while the Marshall–Palmer distribution includes all the measurements in a single distribution. The Marshall–Palmer is seen to be similar to the Joss widespread rain case.

The specific attenuation as given by Equation (4-8) can be now expressed as

$$\alpha \left(\frac{\text{dB}}{\text{km}}\right) = 4.343 N_0 \int Q_t(r, \lambda, m)\, e^{-\Lambda r}\, dr \qquad (4\text{-}13)$$

where Q_t is obtained from Equations (4-9) or (4-11). The above integral equation can be solved numerically for specified values of refractive index, frequency, and drop size distribution. Figure 4-2 shows the specific attenuation from 1 to 1000 GHz at four specified rain rates obtained by a series expansion solution of the Mie scattering coefficients for the three drop size distributions described above. The calculations, made at a rain temperature of 20°C, indicate that the attenuation increases rapidly as a function of frequency up to about 60 GHz, then levels off and drops slightly at higher frequencies. There is little difference in the results for the three distributions below 30 GHz, while at the higher frequencies the Marshall–Palmer and Joss distributions give higher values because of the greater number of small size drops resulting from the exponential representation of these distributions.

The total rain attenuation for a given path is then obtained by integrating the specific attenuation over the total path L, as given by Equations (4-2) and (4-13), i.e.,

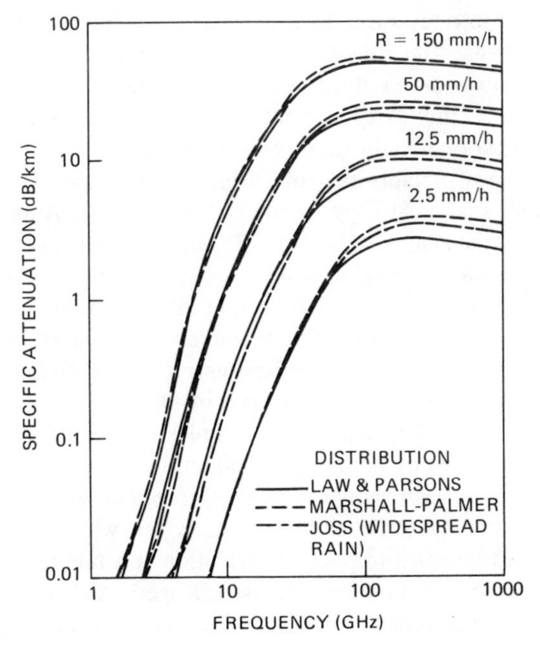

Figure 4-2. Specific Attenuation of rain for three drop size distributions. *R* is the rain rate in mm/h.

$$A(\text{dB}) = 4.343 \int_0^L \left[N_0 \int Q_t \, e^{-\Lambda r} \, dr \right] dx \tag{4-14}$$

where the integration over x is taken over the extent of the rain volume in the direction of propagation. In general both Q_t and the drop size distribution will vary along the path and these variabilities must be included in the integration process. A determination of the variations along the propagation path is very difficult to obtain, particularly for slant paths to an orbiting satellite. We shall see later that these variations must be approximated or treated statistically for the development of useful rain attenuation prediction models.

4.1.1. Attenuation and Rain Rate

When measurements of rain attenuation on a terrestrial path were compared with the rain rate measured on the path, it was observed that the specific attenuation (dB/km) could be well approximated by

$$\alpha \left(\frac{\text{dB}}{\text{km}} \right) = aR^b \tag{4-15}$$

where R is the rain rate in mm/h, and a and b are frequency and temperature dependent constants. The constants a and b represent the complex behavior of the complete representation of the specific attenuation as given by Equation (4-13). This relatively simple expression for attenuation and rain rate was observed by early investigators directly from measurements [4.1, 4.3]; however, several recent studies, most notably that of Olsen, Rogers, and Hodge [4.10], have demonstrated an analytical basis for the aR^b expression. Appendix C presents a development of the analytical basis for the aR^b representation described above.

The use of the aR^b expression is included in virtually all present models for the prediction of path attenuation from rain rate, and several sources of tabulations for the a and b coefficients are available [4.4, 4.10, 4.11]. Table 4-2 presents a listing of the a and b coefficients for several frequencies of interest for satellite communications, as developed by Olsen, Rogers, and Hodge for a 0°C rain temperature. Also listed are examples of specific attenuation at each frequency for rain rates of 10, 50, and 100 mm/h.

The aR^b approximation shows excellent agreement with direct Mie calculations of specific attenuation, as shown in Figure 4-3, which shows a comparison for the Laws and Parsons drop size distribution at a rain temperature of 0°C. The a and b coefficients are those of Olsen, Rogers, and Hodge.

The above development involves rain attenuation caused by spherical rain drops, therefore the results are independent of the polarization sense of the transmitted radiowave. The a and b coefficients for nonspherical drops, which are more representative of a true rain drop, can also be calculated by applying techniques similar to the Mie calculations for a spherical drop. The results are presented in Table 4-3, where the coefficients for a horizontally polarized wave

Table 4-2. Examples of Attenuation Coefficients and Specific Attenuation for Rain. (Marshall–Palmer Drop Size Distribution; Rain Temperature 0°C.)

Frequency (GHz)	Coefficients		Specific Attenuation (dB/km)		
	a	b	$R = 10$	$R = 50$	$R = 100$
2	0.000345	0.891	0.003	0.1011	0.021
4	0.00147	1.016	0.015	0.078	0.158
6	0.00371	1.124	0.049	0.30	0.657
12	0.0215	1.136	0.29	1.83	4.02
15	0.0368	1.118	0.48	2.92	6.34
20	0.0719	1.097	0.90	5.25	11.24
30	0.186	1.043	2.05	11.0	22.7
40	0.362	0.972	3.39	16.2	31.8
94	1.402	0.744	7.78	25.8	43.1

R = rain rate in mm/h.

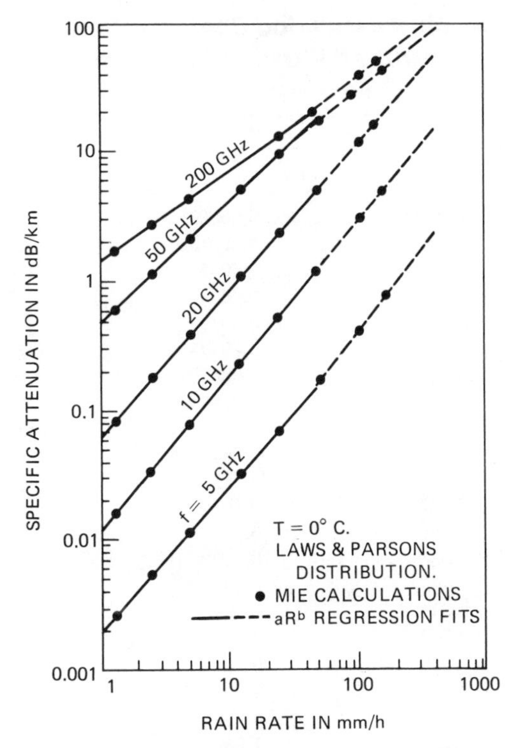

Figure 4-3. Comparison of specific attenuation from Mie calculations and the aR^b approximation.

(a_h, b_h) and for a vertically polarized wave (a_v, b_v) are listed. The coefficients are for a Laws and Parsons drop size distribution at a temperature of 20°C [4.1]. Oblate spheroidal drops with a vertical axis of rotation were assumed.*

The coefficients for a *circularly polarized* wave are calculated from the values in Table 4-3 by,

$$a_c = \frac{a_h + a_v}{2} \tag{4-16}$$

and

$$b_c = \frac{a_h b_h + a_v b_v}{2a_c} \tag{4-17}$$

*A more complete discussion of rain attenuation caused by nonspherical drops is presented in Chapter 6 on depolarization effects.

Table 4-3. Specific Attenuation Coefficients for Rain Attenuation Calculations. (Laws and Parsons Drop Size Distribuiton; Rain Temperature 20°C.)

Frequency (GHz)	a_h	a_v	b_h	b_v
1	0.0000387	0.0000352	0.912	0.880
2	0.000154	0.000138	0.963	0.923
4	0.000650	0.000591	1.121	1.075
6	0.00175	0.00155	1.308	1.265
7	0.00301	0.00265	1.332	1.312
8	0.00454	0.00395	1.327	1.310
10	0.0101	0.00887	1.276	1.264
12	0.0188	0.0168	1.217	1.200
15	0.0367	0.0335	1.154	1.128
20	0.0751	0.0691	1.099	1.065
25	0.124	0.113	1.061	1.030
30	0.187	0.167	1.021	1.000
35	0.263	0.233	0.979	0.963
40	0.350	0.310	0.939	0.929
45	0.442	0.393	0.903	0.897
50	0.536	0.479	0.873	0.868
60	0.707	0.642	0.826	0.824
70	0.851	0.784	0.793	0.793
80	0.975	0.906	0.769	0.769
90	1.06	0.999	0.753	0.754
100	1.12	1.06	0.743	0.744
120	1.18	1.13	0.731	0.732
150	1.31	1.27	0.710	0.711
200	1.45	1.42	0.689	0.690
300	1.36	1.35	0.688	0.689
400	1.32	1.31	0.683	0.684

NOTE: To interpolate between the given frequencies, use logarithmic scale for frequency and *a*, and linear scale for *b*.
SOURCE: CCIR [4.12].

The coefficients for a *linearly polarized* wave, other than horizontal or vertical, are calculated from the values in Table 4-3 by

$$a_\delta = \tfrac{1}{2}[a_h + a_v + (a_h - a_v)\cos^2\theta\cos 2\delta] \tag{4-18}$$

and

$$b_\delta = \frac{1}{2a_\delta}[a_h b_h + a_v b_v + (a_h b_h - a_v b_v)\cos^2\theta\cos 2\delta] \tag{4-19}$$

where θ is the path elevation angle and δ is the polarization tilt angle relative to horizontal.

The aR^b relation is utilized in the development of several rain attenuation prediction models in Chapter 5.

4.1.2. Slant Path and Elevation Angle Dependence

The total rain attenuation experienced on a slant path at an elevation angle θ is determined from the specific attenuation α by

$$A_\theta = \frac{L}{\sin\theta}\alpha = \frac{L}{\sin\theta}aR^b, \quad \text{dB} \tag{4-20}$$

where L is the extent of the rain volume in the direction of propagation. If the rain rate is not constant over the total path length L, which is usually the case, the total attenuation can be found by summing the incremental attenuation for each portion of the path, i.e.,

$$A_\theta = \frac{L_1}{\sin\theta}aR_1^b + \frac{L_2}{\sin\theta}aR_2^b + \cdots = \frac{a}{\sin\theta}\sum_{i=1}^{N} L_i R_i^b \tag{4-21}$$

The path attenuation at another elevation angle ϕ can be determined from the path attenuation at elevation angle θ from

$$A_\phi = \frac{\sin\theta}{\sin\phi}A_\theta \tag{4-22}$$

where the rain is assumed to be horizontally stratified over the interaction region in the vicinity of the ground terminal. The above results are valid for elevation angles above about 10°, where the earth's curvature introduces negligible errors in the surface path projections.

The major problem in the estimation of slant path attenuation is the determination of the extent of the path length L, and the rain rate profile along that path. A large portion of the significant research accomplished on the effects of rain on satellite communications links has been involved with the determination of techniques and models to characterize the slant path from measureable quantities such as the surface rain rate and the 0°C isotherm height. The results of these modeling techniques will be summarized in Chapter 5. A review of measured rain attenuation statistics acquired over the past several years on earth satellite links is presented in the next section.

4.2. RAIN ATTENUATION MEASUREMENTS

Extensive experimental research has been performed on the direct measurement of rain effects on earth-space paths, beginning in the late 1960s, with the availability of propagation beacons on geostationary satellites [4.13, 4.14]. The major satellites included the NASA Applications Technology Satellites (ATS-5, ATS-6), the Canadian/U.S. Communications Technology Satellite (CTS), the domestic U.S. COMSTAR communcations satellites, the ETS-II, CS, and BSE satellites of Japan, SIRIO (Italy), and OTS (European Space Agency). Table 4-4 lists the launch date, orbit location and frequencies for each of the satellites.

Long term attenuation data are most effectively presented in the form of cumulative distributions, usually on an annual or monthly basis, for communications link design. In this format, link reliability (availability) or exceedance (outage) can be determined and rain margins can be designed into the system to achieve the desired link performance. Figure 4-4, for example, shows cumulative attenuation distributions for three consecutive twelve month periods measured at Greenbelt, Maryland at 11.7 GHz with the CTS satellite beacon. The elevation angle to the satellite was 29 degrees, and the ground antenna diameter was 4.6 meters. The distributions provide an estimate of the rain mar-

Table 4-4. Satellites Used for Rain Attenuation Measurements on Earth–Space Paths.

Satellite	Launch Date	Orbit Location	Frequencies (GHz) Uplink	Frequencies (GHz) Downlink
ATS-5	8–69	104°W	31.6 GHz	15.3 GHz
ATS-6	5–74	95°W 35°E (6-75) 105°W (11-76)	13.2 17.7	20 30
CTS (CANADA)	1–76	116°W	14–14.3	11.7–12.2
COMSTAR	5–76 (D1) 7–76 (D2) 6–78 (D3) 2–81 (D4)	128°W 95° W (8-78) 95°W 87°W 127°W	—	19 28 ± .264
ETS-II (JAPAN)	2–77	130°E	—	1.7 11.5 34.5
SIRIO (ITALY)	8–77	15°W	11.6	17.4
CS (JAPAN)	12–77	135°E	27.55–30.55	17.75–20.25
BSE (JAPAN)	4–78	110°E	14–14.5	11.7–12.2
OTS (ESA)	5–78	10°E	14.2–14.5	11.5–11.8

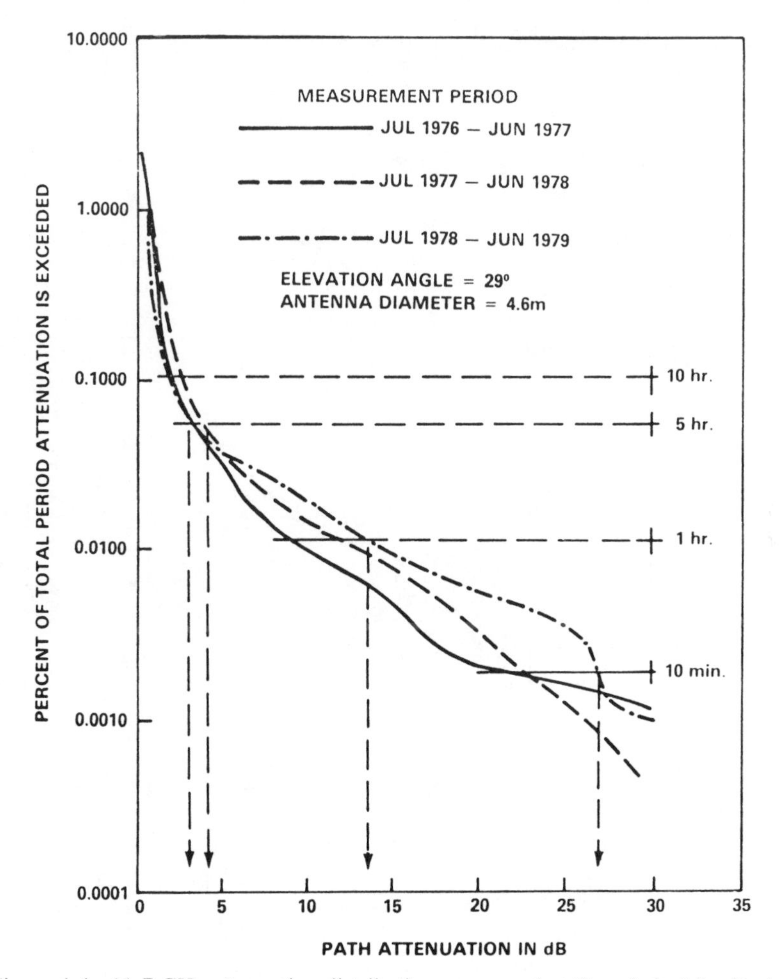

Figure 4-4. 11.7 GHz attenuation distributions measured at Greenbelt, Maryland.

gins required for a given link reliability, usually expressed in percent, or equivalently for a given link reliability, expressed in hours or minutes. For a 1 hour per year (0.011%) link outage (equivalent to a 99.989% link availability), a 13.5 dB margin would have been required for the worst 12 month period of the three years measured. Similarly, margins of 3 dB, 4.5 dB, and 27 dB would have been required for 10 hours, 5 hours, or 10 minutes per year link outage, respectively. A link with a 1 hour per year outage is considered a high quality link, similar to what is specified for the national telephone system.

Table 4-5 summarizes the results of annual 11 GHz measurements observed in the United States, Europe, and Japan utilizing the CTS, SIRIO, BSE, and

Table 4-5. Summary of 11 GHz Annual Attenuation Measurements.

Location	Frequency (GHz) Polarization	Elevation Angle	Time Period	Attenuation (dB) for Given %			
				1%	0.1%	0.01%	0.001%
Waltham, USA	11.7 circ.	24°	February 1977–January 1978	<1	2.5	10.5	—
			February 1978–January 1979				
Holmdel, USA	11.7 circ.	27°	June 1976–June 1977	<1	3	13.5	—
			June 1977–June 1978	<1	3	13.5	—
			June 1978–June 1979	<1	2.5	9.2	29
Greenbelt, USA	11.7 circ.	29°	July 1976–June 1977	<1	1.8	8.8	>30
			July 1977–June 1978	<1	2.1	12	26.4
			July 1978–June 1979	<1	1.8	14	29.2
Blacksburg, USA	11.7 circ.	33°	January 1978–December 1978	2	3.7	6.8	13
			January 1977–December 1977	2	4	13	24
Austin, USA	11.7 circ.	49°	February 1978–January 1979	<1	3	13	23
Munich, Germany	11.6 lin.	29°	January 1978–December 1978 (91.2% coverage)	3	6.8	—	
Fucino, Italy	11.6 circ.	33°	January 1978–December 1980	<1	1.7	4.8	12
Lario, Italy	11.6 circ.	32°	January 1978–December 1980	<1	2.7	8	16
Spino d'Adda, Italy	11.6 circ.	32°	October 1978–September 1980	<1	2.7	6.8	12
Gometz-la-Ville, France	11.6 circ.	32°	November 1977–November 1978	1.3	3.7	6.3	8.5
	11.8 circ.	33.6°	January 1979–November 1979	<1	2.5	5	10.5
Nederhorst Den Berg, Netherlands	11.6	27.5°	August 1975–October 1975 April 1976–June 1977 (8100 hours)	0.8	1.5	3.2	6

Kjeller, Norway	11.6 lin.	22°	May 1980–September 1980	0.8	2.1	6.9	12.6
Bergen, Norway	11.6 lin.	21°	May 1979–September 1979	1.7	3.4	6.4	10.1
Trondheim, Norway	11.6 lin.	18°	May 1979–July 1979	0.9	2.0	4.1	—
Eik, Norway	11.6 lin.	23°	April 1980–June 1980	1.0	2.0	4.2	—
Leeheim, Germany (Federal Republic of)	11.6 lin.	32.9°	January 1979–December 1980	<1	1.1	5.7	13.8
Slough, UK	11.8 circ.	30.3°	July 1978–August 1980	—	<3	4.3	16
Slough, UK	11.6 circ.	29.5°	September 1977–August 1980	—	<3	3	9
Martlesham, UK	11.8 lin.	29.9°	July 1978–June 1980	—	<3	3.7	7.2
Martlesham, UK	11.8 circ.	29.9°	July 1978–June 1980	—	<3	5.6	10.4
Kashima, Japan	11.7	37°	August 1978–July 1980	<1	2	7	
	11.5	47°	May 1977–April 1978	<1	2.5	6.2	
Wakkanai, Japan	12.1	29°	1 year	<1	1.7	2.9	8.0
Yamagawa, Japan	12.1	47°	July 1979–June 1980	1	3	8	
Kasennuma, Japan	12.1	34°	1 year	0.8	2.4	5.9	10.6
Osaka, Japan	12.1	41°	1 year	1.7	5.6	15.1	—
Owase, Japan	12.1	42°	1 year	1.6	11.5	>20	
Matsue, Japan (0°E, 20 m)	12.1	42°	1 year	1.3	4.3	13.3	
Ogasawara, Japan	12.1	43°	1 year	1.0	3.1	7.9	11.0
Ashizuyi, Japan	12.1	45°	1 year	1.9	5.0	10.2	13.6
Izuhara, Japan	12.1	45°	1 year	0.7	2.1	4.9	8.6
Minamidaito, Japan	12.1	52°	1 year	1.4	6.0	16.7	>20
Yonaguni, Japan	12.1	58°	1 year	2.7	8.1	14.3	16.7

SOURCES: Ippolito [4.14], CCIR [4.15].

ETS-II satellite beacons at 11.7, 11.6, 11.7, and 11.5 GHz, respectively. The U.S. locations tend to exhibit a more severe attenuation than the European locations for a given percent outage, which is to be expected since the rain conditions are more severe in U.S. regions, particularly for thunderstorm occurrence periods, which account for most of the attenuation observed above about 0.05% outage. The data from Japan tends to fall somewhere between the U.S. and European results. The mean attenuation for a 0.01% outage (53 minutes per year) is 11.2 dB for the U.S. sites, 4 dB for the European sites, and 5.8 dB for Japan. Some variability between locations is to be expected because of the different observation periods and elevation angles, however, the data does exhibit a fairly consistent trend and gives a good indication of what could be expected on operational communications links.

The five U.S. locations have acquired over 25 station-years of 11.7 GHz satellite attenuation data and the long term attenuation distributions are shown in Figure 4-5. The multi-year distributions demonstrate the dual mode characteristic of rain induced attenuation, which results in two different slopes on the distribution. The attenuation values below about 0.05% per year are caused by stratiform rain and the higher attenuation tails of the distribution are produced by the more severe convective rain usually associated with thunderstorm activity. The resulting rain margins for the four stations which have nearly identical elevation angles around 30 degrees are 2.5 ± 1 dB, 11.2 ± 2.2 dB, and 14.5 ± 3 dB for link reliabilities of 99.9, 99.99, and 99.995%, respectively.

A large data base of rain attenuation measurements in the 20 and 30 GHz frequency bands is available in the United States, beginning with measurements in 1974 using the ATS-6 satellite. Tables 4-6 and 4-7 present listings of annual attenuation statistics at 20 and 30 GHz, respectively, for several U.S. locations and elevation angles ranging from 18.5 to 52 degrees. The attenuation is, as expected, more severe with increasing frequency, reaching very high levels at 30 GHz for outages of 0.05% and less. The mean attenuation values at 0.05% outage are 2.7 dB at 20 GHz and 5.7 dB at 30 GHz. Mean values for 0.1% outage are 8.7 dB at 20 GHz and 15.8 dB at 30 GHz.

The experimental measurements of rain attenuation have to be evaluated with an awareness of the wide variabilities that exist in weather conditions, elevation angle, and measurement techniques. Even with these variations, however, some general observations on the magnitude of rain attenuation on space communications can be made. From the nearly 40 station-years of measurements in the temperate continental regions of the United States, projections can be developed for the rain margins required in each frequency band for a specified link reliability. Table 4-8 presents this information for a range of typical link reliabilities, with the specified hours per year outage. The margins listed are based on an upper bound margin of at least one standard deviation above the mean value calculated for all of the available data in each frequency band.

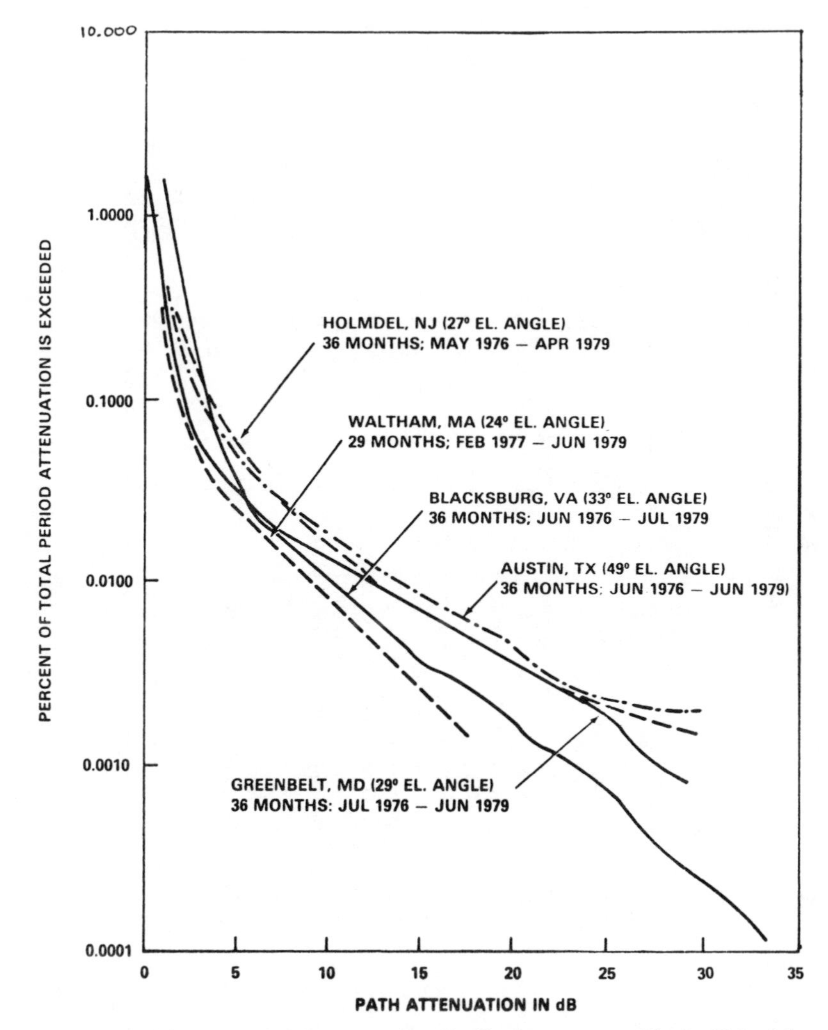

Figure 4-5. Multi-year 11.7 GHz attenuation distributions measured in the United States.

In operational space communications systems, power margins of 6–10 dB can be achieved with a reasonable system design, but margins above 10–15 dB can be costly and difficult to achieve. From the measured results it can be seen that a 99.5% link reliability is a reasonable value for any of the three frequency bands. Reliabilities of 99.95% or better at 20 GHz or 99.9% or better at 30 GHz, however, will require other techniques, such as site diversity or spot beams, in addition to the basic power margin designed into the link.

A summary of rain attenuation prediction methods is presented in Chapter 5,

Table 4-6. Summary of 20 GHz Annual Attenuation Measurements.

Location	Frequency (GHz) Polarization	Elevation Angle	Time Period	Attenuation (dB) for Given %			
				1%	0.1%	0.01%	0.001%
Holmdel, USA	19 lin.	18.5°	June 1976–June 1977	2.3	12	>40	—
Grant Park, USA	19 lin.	27.3°	July 1976–June 1977	—	9.5	>30	—
Palmetto, USA	19 lin.	29.9°	June 1976–July 1977	—	10	>28	—
Waltham, USA	19 lin.	35.3°	January 1978–December 1978[1]	<2	3.5	29	—
Holmdel, USA	19 lin.	38.6°	May 1977–May 1978	1	6.5	22	>44
Grant Park, USA	19 lin.	41.8°	August 1977–August 1978	—	9	>30	—
Blacksburg, USA	19 lin.	45°	January 1978–December 1978	1.7	4.9	11.3	24.5
			January 1979–December 1979[2]	2	5	13.5	25
Rosman, USA	15.3 lin.	45°	January 1974–December 1974	2	10	—	—
Palmetto, USA	19 lin.	49.5°	August 1977–August 1978	—	9	>30	—
Austin, USA	19 lin.	52°	October 1978–October 1979	1	7.2	23	—
Tampa, USA	19 lin.	54.5°	January 1978–December 1978	—	21	—	—
			January 1979–December 1979	<1	>30	—	—
Kashima, Japan	19.5	48°	April 1978–March 1980	2	6	17	—
Yokosuka, Japan	19.5	48°	April 1978–March 1980	3	7	27	—
Yokohama, Japan	19.5	48°	April 1978–March 1980	4	8	16	—
Wakkanai, Japan	19.5	37°	1 year	<1	2.9	8.4	14.3
Yamagawa, Japan	19.5	53°	1 year	2	10	17	—
Sendai, Japan	19.5	45°	April 1979–March 1980	3	7	18	—

[1]September excluded.
[2]February excluded.
SOURCES: Ippolito [4.14], CCIR [4.15].

Table 4-7. Summary of 30 GHz Annual Attenuation Measurements.

Location	Frequency (GHz) Polarization	Elevation Angle	Time Period	Attenuation (dB) for Given % 1%	0.1%	0.01%	0.001%
Grant Park, USA	28.6 lin.	27.3°	July 1976–June 1977	—	17	—	—
Palmetto, USA	28.6 lin.	29.9°	June 1976–July 1977	—	20	—	—
Waltham, USA	28.6 lin.	35.3°	January 1978–December 1978[1]	<2	7.5	>30	—
Holmdel, USA	28.6 lin.	38.6°	May 1977–May 1978	3.5	14.5	44	—
Wallops Island, USA	28.6 lin.	41.6°	April 1977–Marcy 1978	2.3	12.5	—	—
Grant Park, USA	28.6 lin.	41.8°	August 1977–August 1978	—	20	—	—
Blacksburg, USA	28.6 lin.	45°	January 1978–December 1978	5	11.8	24	—
			January 1979–December 1979[2]	4	8.5	25	—
Rosman, USA	31.6 lin.	45°	January 1974–December 1974	3.5	22	>25	—
Palmetto, USA	28.6 lin.	49.5°	August 1977–August 1978	—	18	—	—
Austin, USA	28.6 lin.	52°	October 1978–October 1979	1.5	17	—	—
Kashima, Japan	34.5	47°	May 1977–April 1978	5 (5)	19.5 (18)	—	—

[1]September excluded.
[2]February excluded.
SOURCES: Ippolito [4.14], CCIR [4.15].

Table 4-8. Rain Margins Required in the United States for a Given Link Reliability, from Direct Satellite Measurements.

Link Reliability %	Hours per Year Outage	Margin (db)		
		11 GHz	20 GHz	30 GHz
99.5	44	1	3	6
99.9	8.8	3	10	20
99.95	4.4	5	20	>30
99.99	0.88	15	>30	—

and the effects of rain attenuation on system performance are discussed further in Chapters 10 and 11, where satellite link performance, reliability, and outage considerations are developed in more detail.

4.3. CLOUD AND FOG ATTENUATION

Although rain is the most significant hydrometeor affecting radiowave propagation, the influence of clouds and fog can also be present on an earth–space path. Clouds and fog generally consist of water droplets of less than 0.1 mm in diameter, while raindrops typically range from 0.1 mm to 10 mm in diameter. Clouds are water droplets, not water vapor; however, the relative humidity is usually near 100% within the cloud. High-level clouds, such as cirrus, are composed of ice crystals which do not contribute substantially to radiowave attenuation but can cause depolarization effects (see Section 6.2).

The average liquid water content of clouds varies widely, ranging from 0.05 to over 2 g/m^3. Peak values exceeding 5 g/m^3 have been observed in large cumulus clouds associated with thunderstorms, but peak values for fair weather cumulus are generally less than 1 g/m^3. Table 4-9 summarizes the concentration, liquid water content, and droplet diameter for a range of typical cloud types [4.16].

The small size of cloud and fog droplets allows the Rayleigh approximation to be employed in the calculation of specific attenuation. This approximation is valid for radiowave frequencies up to about 100 GHz. The procedure for determining the specific attenuation is very similar to that described earlier for the case of rain attenuation. The attenuation coefficient for rain and fog, k_c, employing the Rayleigh approximation, is found as follows [4.17]:

$$k_c = 0.4343 \frac{6\pi}{\lambda} M \operatorname{Im}\left[\frac{m^2 - 1}{m^2 + 2}\right] \qquad (4\text{-}23)$$

where M is the liquid water content (sometimes called density) in g/m^3, λ is

Table 4-9. Observed Characteristics of Typical Cloud Types.

Cloud Type	Concentration (no/cm^3)	Liquid Water (g/m^3)	Average Radius (microns)
Fair-weather cumulus	300	0.15	4.9
Stratocumulus	350	0.16	4.8
Stratus (over land)	464	0.27	5.2
Altostratus	450	0.46	6.2
Stratus (over water)	260	0.49	7.6
Cumulus congestus	207	0.67	9.2
Cumulonimbus	72	0.98	14.8
Nimbostratus	330	0.99	9.0

SOURCE: Slobin [4.16].

the wavelength in mm, and m is the refractive index of the water droplets. The major difference in the resulting equation for cloud/fog attenuation and rain attenuation is that liquid water content, rather than rain rate, is used for the defining parameter which describes the hydrometeor environment.

Figure 4-6 presents the specific attenuation calculated for the range of liquid water content of typical cloud masses, ranging from 0.05 to 2.5 g/m^3. Also shown in the figure for comparison are the specific attenuations for rain at rain rates of 1, 2.5, and 10 mm/h. Typical cumulus clouds range from 2 to 8 km in extent, which is the same order of magnitude expected for the vertical extent of rain on the earth–space path. The comparisons of Figure 4-6 show that cloud attenuation is equivalent to a very light rainfall attenuation. For a rain which exceeds 10 mm/h, the rain attenuation will be the dominant factor in the total attenuation experienced on the path. For example, at 20 GHz, for a 50 mm/h rain of 4 km extent, the attenuation would be 21 dB. A 4 km extent cumulonimbus cloud at 2.5 g/m^3 would result in an added attenuation of 3.5 dB, under the same prevalent conditions.

4.3.1. Cloud Attenuation Prediction Model

A detailed study of the radiowave propagation effects of clouds at various locations in the United States by Slobin resulted in the development of a cloud model which determines cloud attenuation and noise temperature on radiowave links [4.16]. Extensive data on cloud characteristics, such as type, thickness, and coverage, were gathered from twice-daily radiosonde measurements and hourly temperature and relative humidity profiles at stations in the contiguous United States, Alaska, and Hawaii. Regions of statistically "consistent" clouds

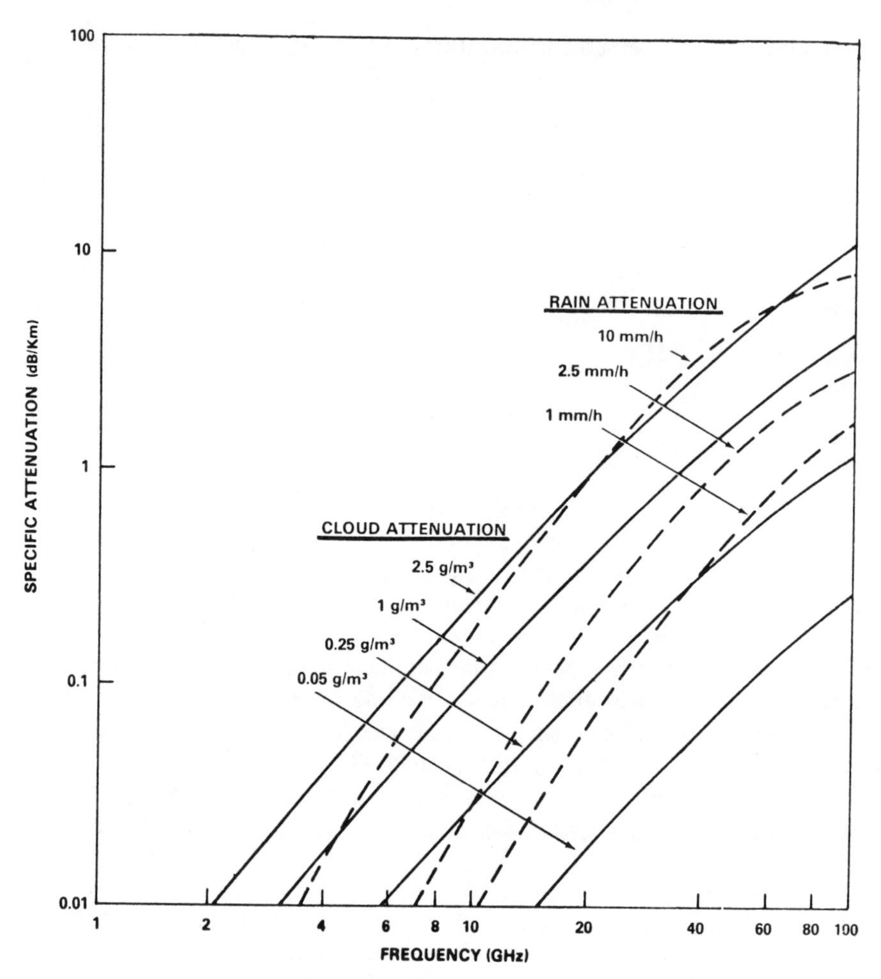

Figure 4-6. Specific attenuation of clouds compared with rain.

were determined, and resulted in the definition of fifteen cloud regions, as shown on the map of Figure 4-7. The regions are identified by three-letter identifiers of the National Weather Service, and are listed in Table 4-10 along with the elevation and 30-year average annual rain statistics for each location.

Twelve cloud types are defined in the Slobin model, based on liquid water content, cloud thickness, and base heights above the surface. Several of the more intense cloud types include two cloud layers, and the combined effects of both are included in the model. Table 4-11 lists seven of the Slobin cloud types, labeled here from light, thin clouds to very heavy clouds, and shows the char-

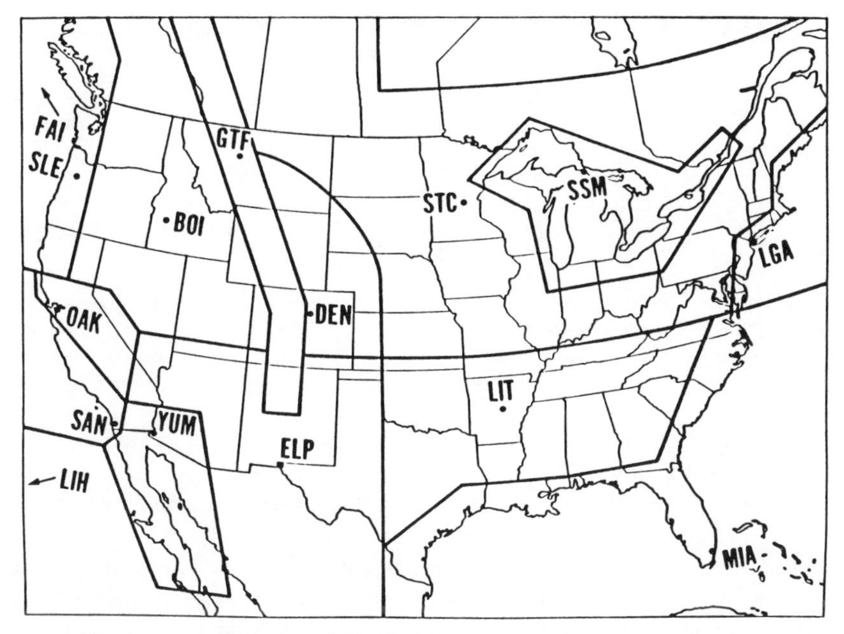

Figure 4-7. Cloud regions of the United States for the Slobin cloud model.

Table 4-10. Location Identifiers and Rainfall Statistics for the 15 Cloud Regions of the United States.

			30-Year Rain Records		
Identifier	Station	Elevation MSL, km	Average (inches)	Wettest Year, (inches)	Driest Year, (inches)
BOI	Boise, Idaho	0.871	11.50	15.77	7.43
DEN	Denver, Colorado	1.661	15.51	21.58	7.51
ELP	El Paso, Texas	1.195	7.77	17.19	4.92
FAI	Fairbanks, Alaska	0.135	11.22	16.62	5.55
GTF	Great Falls, Montana	1.123	14.99	21.59	9.02
LGA	New York (LaGuardia), N.Y.	0.007	40.19	46.39	32.99
LIH	Lihue, Hawaii	0.036	44.18	56.94	21.15
LIT-	Little Rock, Arkansas	0.078	48.52	70.60	28.30
MIA	Miami, Florida	0.004	59.80	89.33	33.75
OAK	Oakland, California	0.006	17.93	26.20	9.90
SAN	San Diego, California	0.009	9.45	19.41	3.41
SLE	Salem, Oregon	0.061	41.08	47.47	29.88
SSM	Sault Sainte Marie, Mich.	0.221	31.70	37.10	25.51
STC	St. Cloud, Minnesota	0.316	25.94	34.60	16.20
YUM	Yuma, Arizona	0.098	2.67	4.84	0.30

Table 4-11. Characteristics of Slobin Model Cloud Types.

Cloud Type	Case No.	Liquid Water (g/m^3)	Lower Cloud Base (km)	Lower Cloud Thickness (km)	Upper Cloud Base (km)	Upper Cloud Thickness (km)
Light, thin	2	0.2	1.0	0.2	—	—
Light	4	0.5	1.0	0.5	—	—
Medium	6	0.5	1.0	1.0	—	—
Heavy I	8	0.5	1.0	1.0	3.0	1.0
Heavy II	10	1.0	1.0	1.0	3.0	1.0
Very Heavy I	11	1.0	1.0	1.5	3.5	1.5
Very Heavy II	12	1.0	1.0	2.0	4.0	2.0

acteristics of each. The case numbers listed in the table correspond to the numbers assigned by Slobin.

The total zenith (90° elevation angle) attenuation was calculated by radiative transfer methods for frequencies from 10 to 50 GHz for each of the cloud types, and the results for several of the frequency bands of interest in space communications are summarized in Table 4-12. The values *include* the clear air gaseous attenuation also. The values at *C*-band and *Ku*-band are less than 1 dB, even for the most intense cloud types.

The Slobin model also developed annual cumulative distributions of cloud attenuation for each of the fifteen cloud regions at fifteen frequencies from 8.5 to 90 GHz. The calculations were based on each hour of input data from the sites, including percent cloud cover, and on an average test year chosen on the basis of rainfall attenuation.

Figure 4-8 shows an example of the distribution of a frequency of 30 GHz for five cloud regions, ranging from very dry, clear Yuma to very wet, cloudy

Table 4-12. Zenith Cloud Attenuation as Predicted by Slobin Cloud Models.

Frequency (GHz)	Light, Thin Cloud	Light Cloud	Medium Cloud	Heavy Clouds I	Heavy Clouds II	Very Heavy Clouds I	Very Heavy Clouds II
6/4	<0.1 dB	<0.1	<0.2	<0.2	<0.2	<0.3	<0.3
14/12	0.1	0.15	0.2	0.3	0.45	0.6	0.9
17	0.2	0.22	0.3	0.45	0.7	1.0	1.4
20	0.25	0.3	0.4	0.6	0.9	1.4	1.8
30	0.3	0.4	0.5	1.0	1.7	2.7	3.9
42	0.7	0.9	1.2	2.1	3.5	5.5	7.9
50	1.5	1.9	2.3	3.6	5.7	8.4	11.7

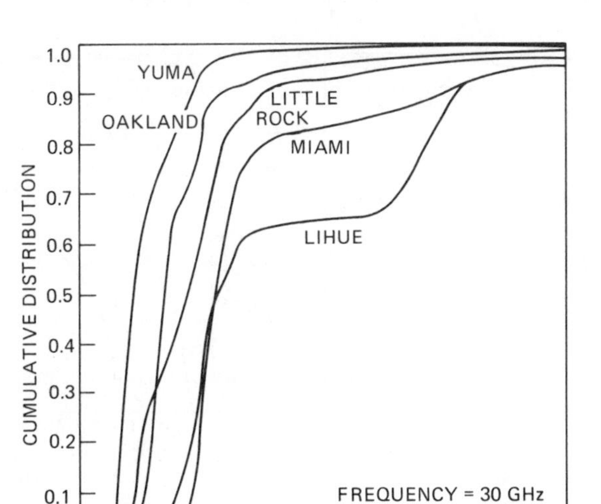

Figure 4-8. Zenith atmospheric attenuation at 30 GHz from the Slobin cloud model.

Lihue. Figure 4-9 (a–d) shows examples of the attenuation distributions for four of the cloud regions, Denver, New York, Miami, and Oakland, at frequencies of 10, 18, 32, 44, and 90 GHz. Plots for all of the cloud regions are available in Reference 4.16.

The distributions give the percent of the time that cloud attenuation is the given value or less. For example, on the Miami plot, the cloud attenuation was 0.6 dB or less for 0.5 (50%) of the time at 32 GHz. Values of attenuation in the distribution range 0–0.5 (0–50%) may be regarded as the range of clear sky effects. The value of attenuation at 0% is the lowest value observed for the test year.

The Slobin model presents a concise and comprehensive model for the evaluation of cloud attenuation on radiowave paths.

4.3.2. Fog Attenuation on Earth–Space Paths

Fog results from the condensation of atmospheric water vapor into suspended air droplets. Moderate fog will have a liquid water content ranging from 0.02 to 0.06 g/m. Heavy fog can range from 0.2 to 0.4 g/m. Fog layers typically extend 50 to 100 meters above the surface, and seldom exceed 150 meters in extent. Because the layers are so thin, fog attenuation on an earth–space path will be very low, and can be considered as negligible in system calculations for frequencies below 100 GHz.

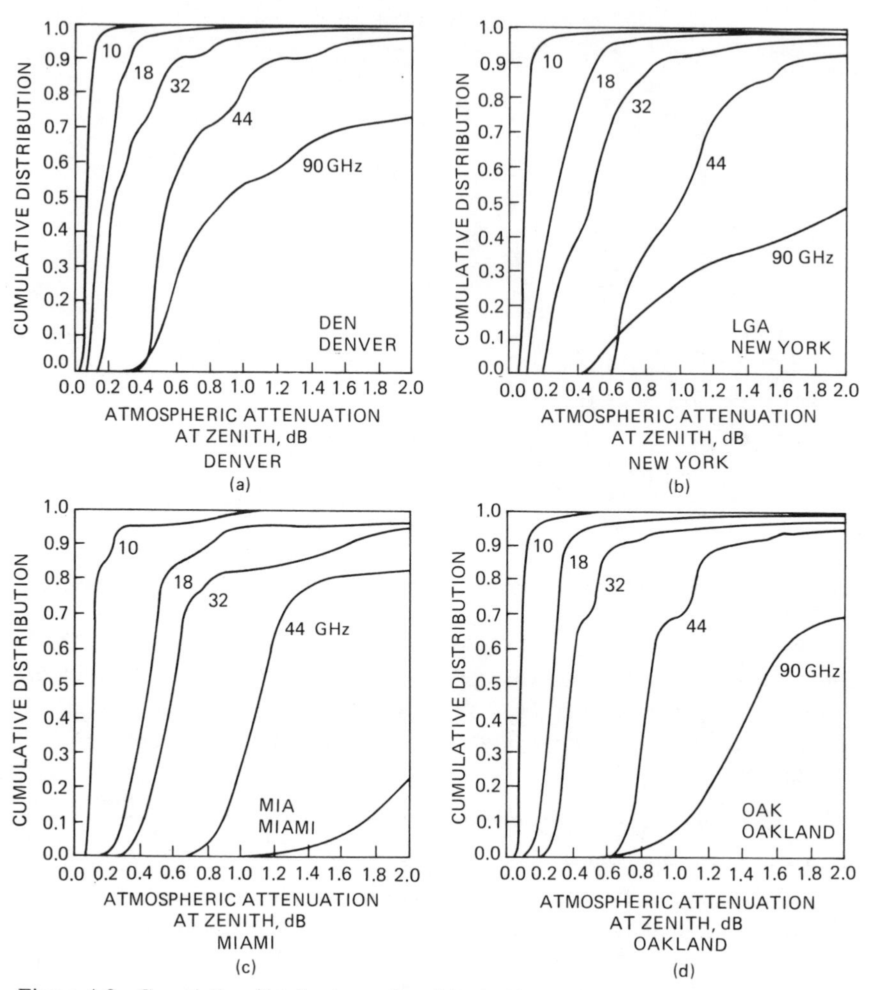

Figure 4-9. Cumulative distributions of zenith cloud attenuation at four locations, from the Slobin model.

REFERENCES

4.1. Ryde, J. W., and Ryde, D., "Attenuation of Centimetere and Millimetre Waves by Rain, Hail, Fogs, and Clouds," Rep. No. 8670, Research Laboratories of the General Electric Co., Wembley, England, 1945.

4.2. Ryde, J. W., "The Attenuation and Radar Echoes Produced at Centimetere Wavelengths by Various Meterological Phenomena," in *Meteorological Factors in Radio Wave Propagation*, the Physical Society, London, pp. 169–188, 1946.

4.3. Gunn, K. L. S., and East, T. W. R., "The Microwave Properties of Precipitation Particles," *Quarterly J. Royal Meteor. Soc.*, Vol. 80, pp. 522–545, 1954.

4.4. Medhurst, R. G., "Rainfall Attenuation of Centimeter Waves: Comparison of Theory and Measurement," *IEEE Trans. Antennas and Propagation*, Vol. AP-13, pp. 550–564, July 1965.
4.5. Mie, G., *Ann. Physik*, Vol. 25, p. 377, 1908.
4.6. Van De Hulst, *Light Scattering by Small Particles*, John Wiley & Sons, New York, 1957.
4.7. Laws, J. O., and Parsons, D. A., "The Relation of Raindrop Size to Intensity," *Trans. Am. Geophys. Union*, Vol. 24, pp. 452–460, 1943.
4.8. Marshall, J. S. and Palmer, W. Mck., "The Distribution of Raindrops with Size," *J. Meteor.*, Vol. 5, pp. 165–166, August 1948.
4.9. Joss, J., and Waldvogel, A., "The Variation of Raindrop Size Distributions at Locarno," *Proc. of the International Conf. on Cloud Physics*, Toronto, Canada, pp. 369–373, 1968.
4.10. Olsen, R. L., Rogers, D. V., and Hodge, D. B., "The Relation in the Calculation of Rain Attenuation," *IEEE Trans. on Antennas and Propagation*, Vol. AP-26, No. 2, pp. 318–329, March 1978.
4.11. Setzer, D. E., "Computed Transmission Through Rain at Microwave and Visible Frequencies," *The Bell System Tech. J.*, Vol. 49, pp. 1878–1892, October 1970.
4.12. CCIR, Report 721-1, "Attenuation by Hydrometeors in Particular Precipitation, and other Atmospheric Particles," in Volume V, *Propagation in Non-Ionized Media*, Recommendations and Reports of the CCIR—1982, International Telecomm. Union, Geneva, pp. 167–181, 1982.
4.13. Kaul, R., Rogers, D. V., and Bremer, J., "A Compendium of Millimeter Wave Propagation Studies Performed by NASA," Operations Research Inc., Report ORI TR 1278, November 1977.
4.14. Ippolito, L. J., "Radio Propagation for Space Communications Systems," *Proc. of the IEEE*, Vol. 69, No. 6, pp. 697–727, June 1981.
4.15. CCIR, Report 564-2, "Propagation Data Required for Space Telecommunications Systems," in Volume V, *Propagation in Non-ionized Media*, Recommendations and Reports of the CCIR—1982, International Telecommunications Union, Geneva, pp. 331–373, 1982.
4.16. Slobin, S. D., "Microwave Noise Temperature and Attenuation of Clouds: Statistics of These Effects at Various Sites in the United States, Alaska, and Hawaii," *Radio Science*, Vol. 17, No. 6, pp. 1443–1454, Nov.-Dec. 1982.
4.17. Battan, L. J., *Radar Meteorology*, University of Chicago Press, Chicago, 1959.

CHAPTER 5
RAIN ATTENUATION PREDICTION METHODS

The evaluation of the effects of rain on a satellite system design requires a detailed knowledge of the attenuation statistics for each ground terminal location at the specific frequency of interest. Direct long-term measurements of rain attenuation for all of the ground terminal locations in an operational network are not practical, therefore modeling and prediction methods must be used to make a best estimate of the expected attenuation for each location. Over the past several years extensive efforts have been undertaken to develop reliable techniques for the prediction of path rain attenuation for a given location and frequency, and the availability of satellite beacon measurements has provided a data base for the validation and refinement of the prediction models. This chapter reviews several of the more promising rain attenuation prediction techniques.

Virtually all of the prediction techniques use surface measured rain rate as the statistical variable and assume the aR^b relationship described earlier to determine rain induced attenuation. In general, the prediction models can be expressed in the form

$$A(\text{dB}) = aR^b\,L(R) \tag{5-1}$$

where R is the rain rate, a and b are the frequency dependent constants described by Equation (4-15), and $L(R)$ is an "effective" path length parameter, usually a specified function of R. The path length parameter $L(R)$ is the coupling function which provides rain attenuation from a specified rain rate distribution, through Equation (5-1). The major difference between the various rain attenuation prediction methods is in the rationale used to develop the path length parameter $L(R)$.

Much of the early work which led to the development of rain attenuation models was focused on empirical methods which described the rain rate distribution as a function of location. One of the earliest attempts to develop a comprehensive rain attenuation prediction technique came through the actions of the International Radio Consultative Committee (CCIR). At the April 1972 Study Group Meetings in Geneva, the CCIR adopted a procedure to estimate the cumultative distribution of rain based on the concept of rain climate zones [5.1]. The procedure, as later modified [5.2], consisted of a global map dividing the land areas of the earth's surface into five climate regions and provided an average annual rain rate distribution for each region. The technique provided very coarse estimates because of the limited number of zones and the lack of beacon data for verification; however, it did serve as the basis for many of the concepts which were employed in later methods.

5.1. RICE–HOLMBERG RAIN MODEL

A global surface rain rate model developed from extensive long term rain rate statistics from over 150 locations throughout the world was developed by Rice and Holmberg [5.3]. The Rice–Holmberg model constructs a rain rate distribution by assuming that the rain structure can be divided into two types, or modes, "thunderstorm rain" and "all other rain." Each mode is modeled by exponential functions and the sum of the two modes produces the total distribution. The percent of an average year for which the rain rate exceeds R mm/h at a medium location is given by,

$$P(R)\% = \frac{M}{87.6}\{0.03\,\beta\, e^{-0.03R} + 0.2(1-\beta)[e^{-0.258R} + 1.86\, e^{-1.63R}]\} \tag{5-2}$$

where M is the average annual rainfall accumulation (in mm), M_1 is the average annual accumulation of *thunderstorm rain* (in mm), $\beta = M_1/M$, and R is the clock minute rain rate, in mm/h.

Global maps for M, M_1, and β are provided in the model, or directly measured data can be used when available. The Rice–Holmberg model, one of the first developed for radiowave propagation studies, has been shown to provide very good agreement with measured data for locations in the United States where long term rain rate data has been measured directly.

5.2. DUTTON–DOUGHERTY ATTENUATION PREDICTION

The global rain rate model of Rice–Holmberg described above does not provide a prediction of rain induced attenuation on a radiowave path. The Rice–Holmberg model, however, was later extended by Dutton and Dougherty to include an attenuation prediction [5.4, 5.5]. The Dutton–Dougherty (DD) Model is based on meteorological considerations of the propagation path. It begins by expressing the clock minute rain rate distribution for an average year in three segments, i.e.,

$$P(R) = \begin{cases} 0.0114(T_{11} + T_{12})\, e^{-R/R_1'}, & R < 5 \text{ mm/h} \\ 0.0114T_{21} \exp\left(-\sqrt[4]{R/R_{21}}\right), & 5 \le R \le 30 \\ 0.0114T_{11}\, e^{-R/\bar{R}_{11}}, & R > 30 \end{cases} \tag{5-3}$$

where T_{11}, T_{21}, R_1', and $\bar{R}_{11}$ are linear combinations of M, β, and $D = 24 + 3M$, determined from the regression equations

$$T_{21} = b_{11}M + b_{21} \pm S_2$$

$$R_1' = b_{31}M + b_{41}\beta + b_{51}D + b_{61} \pm S_3$$

$$\bar{R}_{11} = \frac{\beta M}{T_{11}} = a_{11}M + a_{21}\beta + a_{31}D + a_{41} \pm S_1 \tag{5-4}$$

The DD attenuation distribution prediction is then determined from the rain rate distribution through the following assumptions:

1. A Marshall–Palmer drop size distribution.
2. A storm height distribution of Grantham and Kantor [5.6].
3. Vertical dependence of liquid water content from Dutton [5.7].
4. The specific attenuation coefficients $a(f)$ and $b(f)$ of Crane [5.8].

The DD model provides *total* atmospheric attenuation, that is gaseous, cloud, and rain attenuation. Figure 5-1(a–d) presents examples of the DD prediction curves and compare the results with measured annual attenuation data at four locations and at frequencies of 11.7 and 19 GHz. The DD Model provides a *mean* prediction (50% confidence level) and *upper* and *lower bounds* which are at the 99.5%, 95%, 5%, and 0.5% confidence levels, respectively [5.9]. The solid curves labeled 1 through 5 in Figure 5-1 represent these predictions. The DD Model is seen to provide a very good prediction for the 19 GHz data for Blacksburg. The six years of 11.7 GHz measurements at Greenbelt and Holmdel fall above the mean curves for attenuation values exceeding 6 dB, but remain within the 99.5% curve up to about the 15 dB level. The data from Austin extends above the prediction bounds throughout the measurement range.

The Dutton–Dougherty Model was recently updated by the authors to provide a more flexible and accessible procedure for general use [5.10]. The lower range

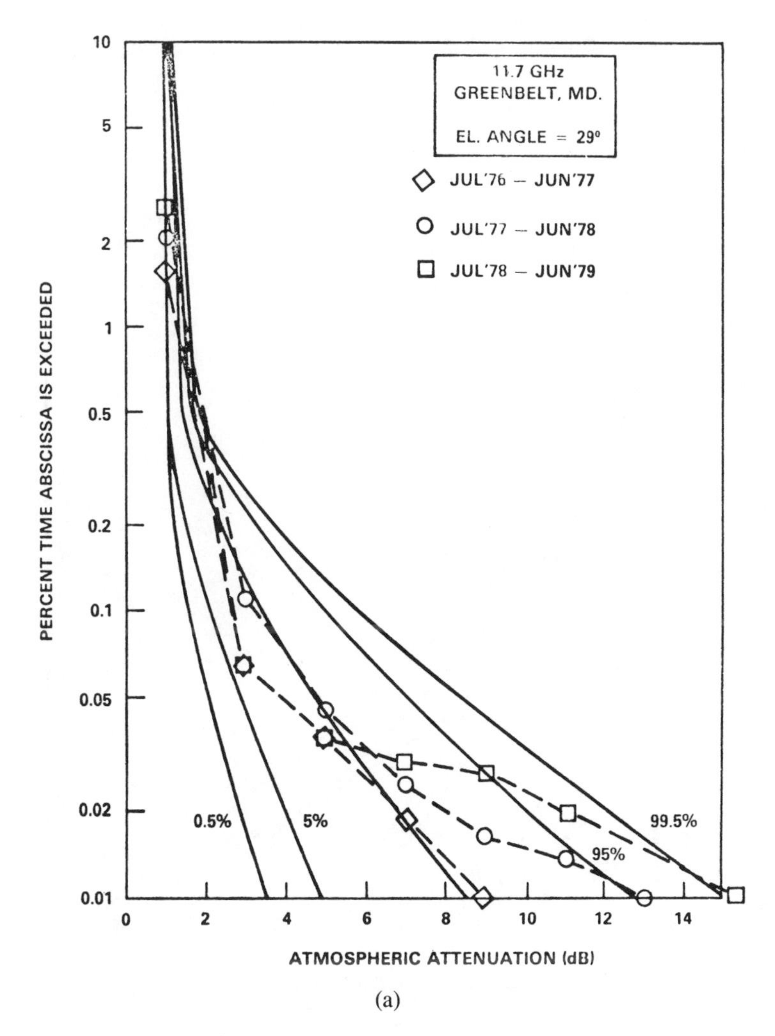

(a)

Figure 5-1. Comparison of the Dutton–Dougherty prediction method with measured data.

of the model prediction was extended to 0.001% (5.3 minutes per year) from the original model limit to 0.01%. The extension was accomplished by employing an analytical method which involved a comparison of the predictions with measured distributions to achieve the best results.

The Dutton–Dougherty Model is now accessible as a user service through the National Telecommunications and Information Agency's Institute for Telecommunications Sciences computer facilities in Boulder, Colorado [5.11].

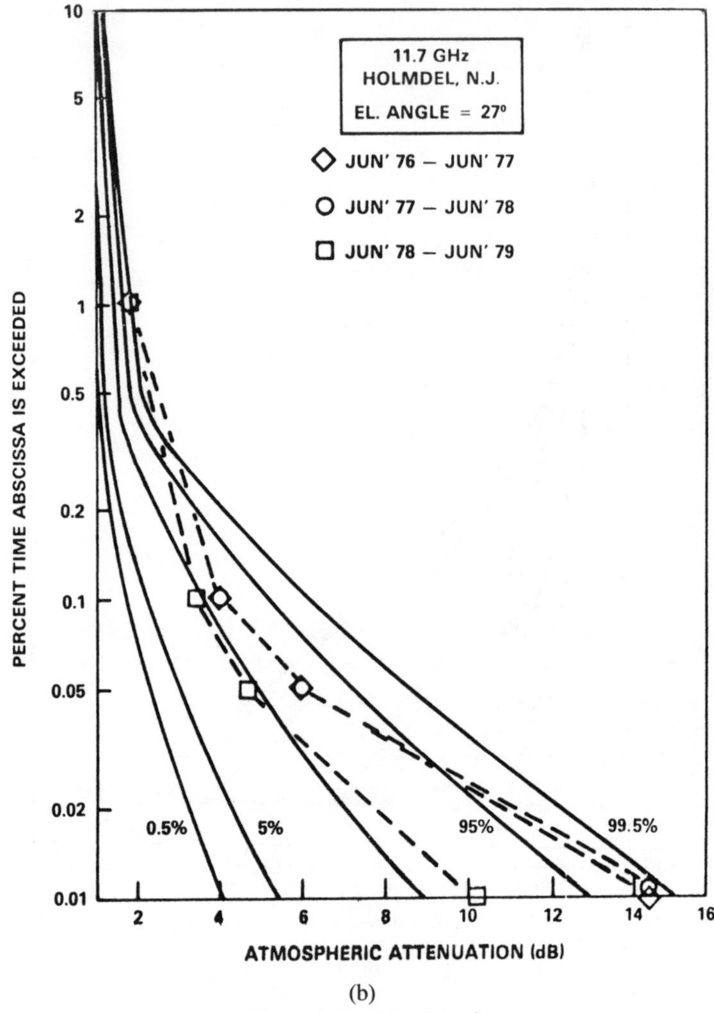

(b)

Figure 5-1. (*Continued*)

5.3. LIN RAIN ATTENUATION MODEL

An empirical method for estimating rain induced outage probabilities on radiowave paths was developed at Bell Laboratories by Lin [5.12, 5.13]. The Lin method is based on a *five-minute* point rain rate rather than the one minute or less rain gage integration times of other prediction techniques. Two arguments are presented to justify the use of the five-minute averaging time. First, the

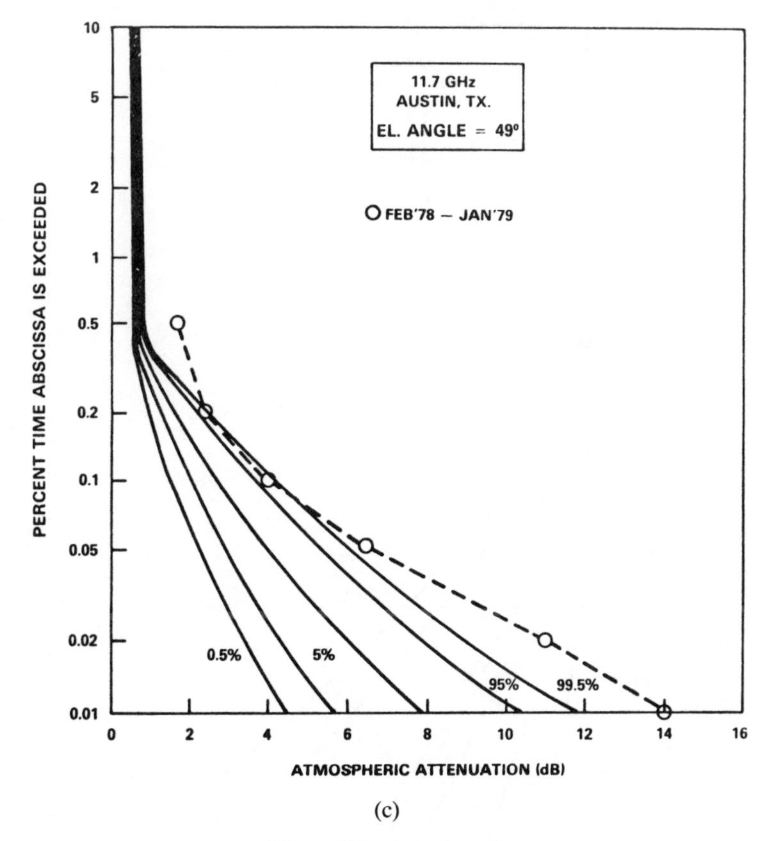

(c)

Figure 5-1. (*Continued*)

available long term rain rate data for the United States, published by the National Weather Service (National Climate Center, Federal Building, Ashville, North Carolina 28801) have a minimum integration times of five minutes, and attempts to estimate shorter times from the original strip chart data produce significant errors. Second, the five minute averaging time effectively produces a path average rain rate distribution from a point rain rate distribution, that is, the spatial average of the rain rate along the radiowave path can be represented by a five minute average of the rain rate at a single point on the path.

The relationship between path attenuation and five minute rain rate was determined empirically from measurements on nine 11 GHz terrestrial paths, 5–43 km in length, at five locations in the United States. The resulting relationship for path attenuation was

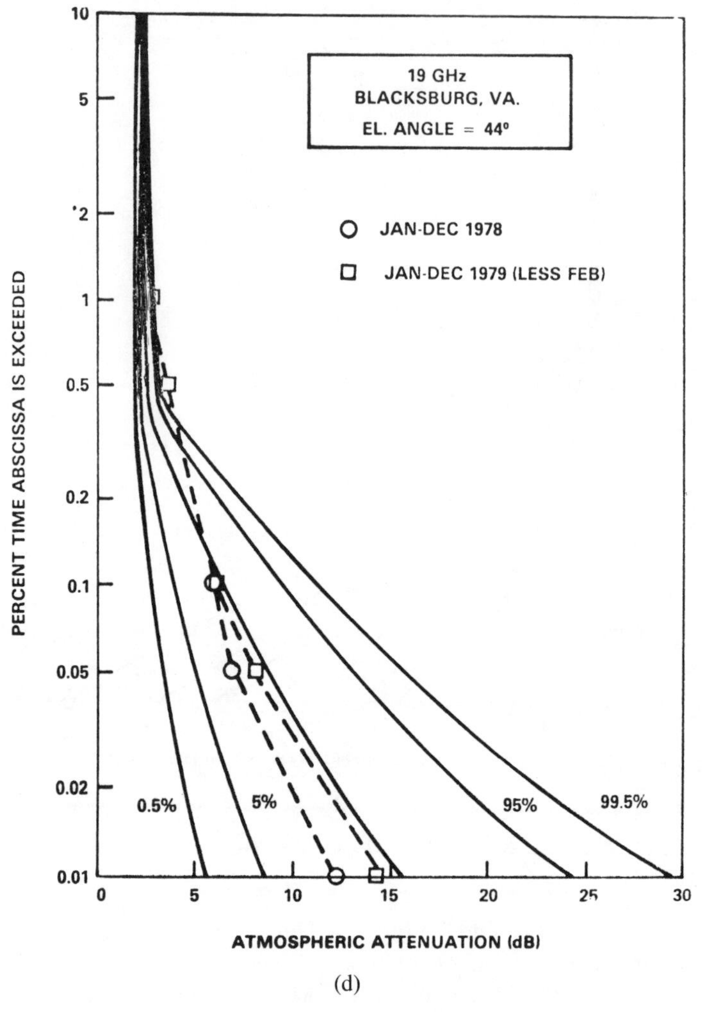

(d)

Figure 5-1. (*Continued*)

$$A_L = aR_5^b \, L \left[\frac{1}{1 + \frac{L}{\bar{L}(R_5)}} \right] \tag{5-5}$$

where R_5 is the five-minute point rain rate, L is the terrestrial path length, and a and b are the frequency dependent constants described by Equation (4-15). The term in brackets is called a *path length correction factor* by Lin, chosen to represent the empirical ratio between the five minute point rain rate and the radiowave path average rain rate at the same probability level [5.13].

The function $\bar{L}(R_5)$ was determined from the terrestrial path measurements as

$$\bar{L}(R_5) \cong \frac{2636}{R_5 - 6.2} \tag{5-6}$$

Lin extended the terrestrial path results to a slant path at an elevation angle θ by geometric considerations similar to those described in Section 4.1.2 to obtain

$$L = \frac{H - G}{\sin \theta} \tag{5-7}$$

where H is the long term average 0° isotherm height, assumed by Lin to be 4 km for Eastern United States locations, and G is the ground terminal elevation (height) above sea level.

The resulting prediction equation for slant path rain attenuation is then

$$A_\theta(\text{dB}) = aR_5^b \left[\frac{2636}{R_5 - 6.2 + \dfrac{2636 \sin \theta}{4 - G}} \right] \tag{5-8}$$

The term in brackets corresponds to the effective path length parameter for the Lin model, as defined by Equation (5-1).

Figure 5-2(a–d) presents a comparison of the attenuation prediction model as calculated by Lin with measured attenuation statistics for four locations in the Eastern United States at frequencies from 11.7 to 28 GHz. The prediction curves are seen to give very good agreement with the measured distributions throughout the range of comparison. The Lin model is limited in its general applicability, however, since it requires five minute average rain distributions, and it is structured primarily for Eastern United States climates.

5.4. CRANE GLOBAL RAIN ATTENUATION MODEL

The first widely published model which provided a self-contained prediction procedure for global application was developed by R. K. Crane [5.13, 5.14]. The Crane global model is based on the use of geophysical data to determine the surface point rain rate, point-to-path variations in rain rate, and the height

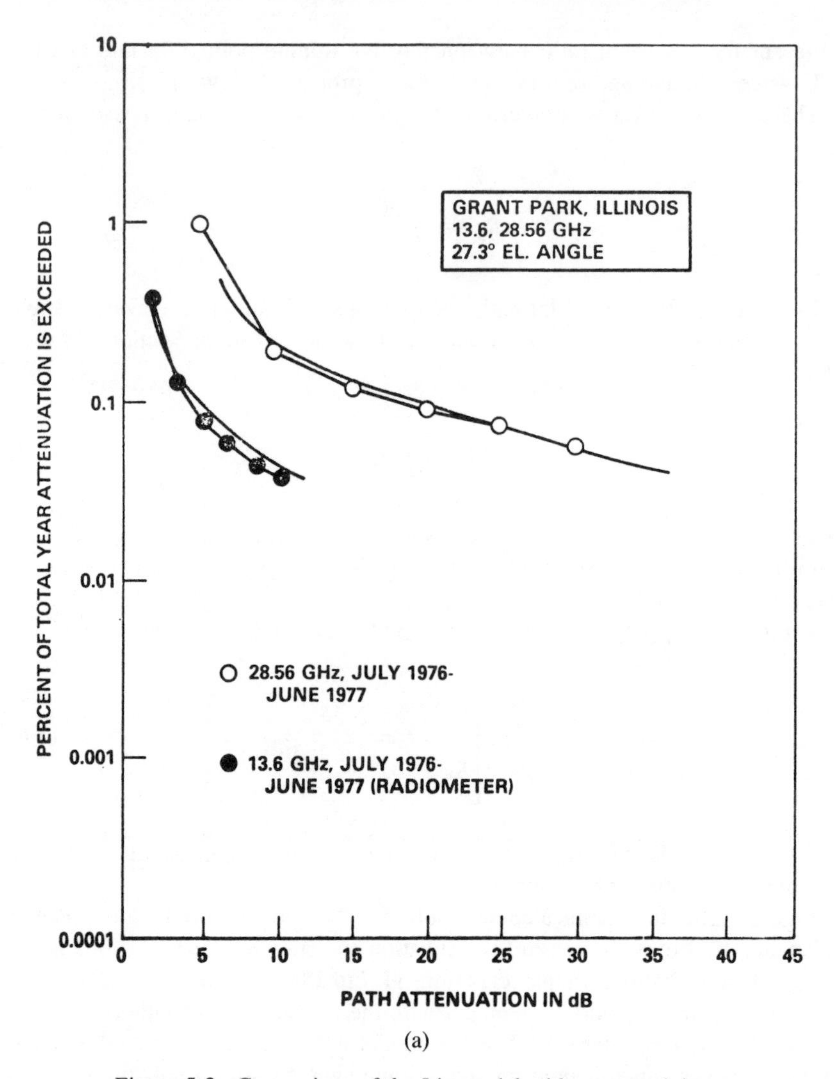

(a)

Figure 5-2. Comparison of the Lin model with measured data.

dependency of attenuation, given the surface point rain rate or the percentage of the year the attenuation value is exceeded. The model also provides estimates of the expected year-to-year and station-to-station variations of the attenuation prediction for a given percent of the year.

Surface point rain rate data for the United States and global sources were used to produce eight rain rate climate regions for the earth's land and water surface areas. The boundaries for each climate region were adjusted for ex-

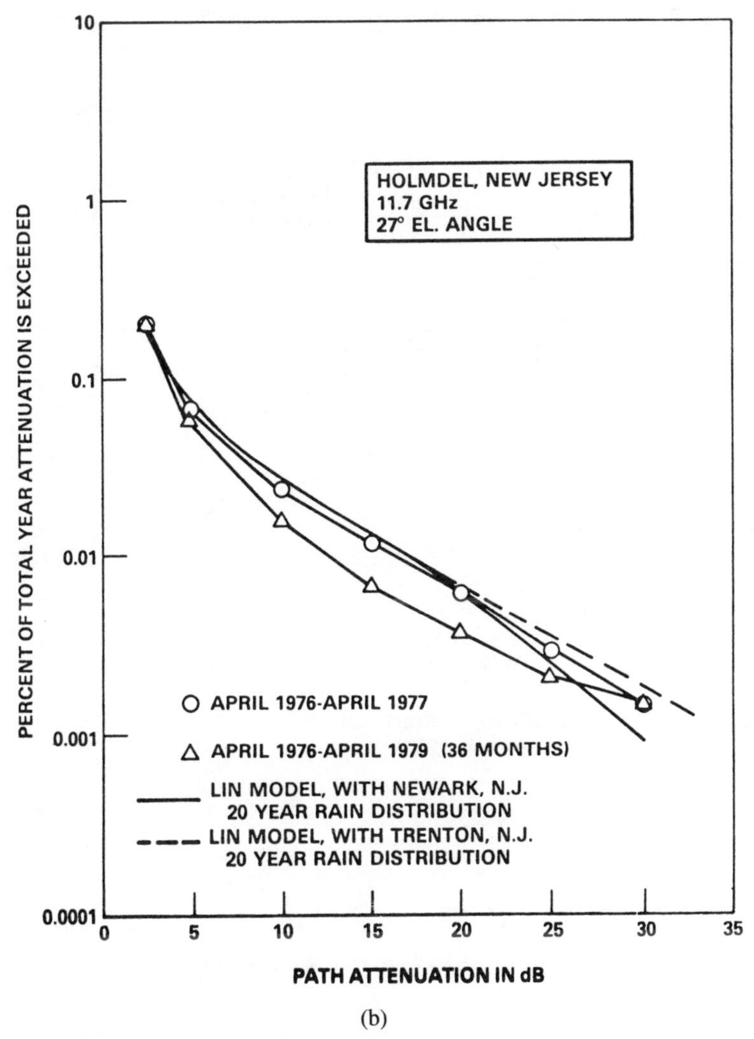

(b)

Figure 5-2. *(Continued)*

pected variations in terrain, storm type, storm motion, and atmospheric circulation. Figure 5-3 shows the resulting global map of the rain rate climate regions, including the ocean areas. The rain climate regions are divided into polar, temperate, subtropical, and tropical, then further subdivided as shown on the figure. As more measured data became available in North America and Europe, regions B and D were further subdivided into regions B_1, B_2, and D_1, D_2, D_3. Figures 5-4 and 5-5 show the resulting rain climate regions for North America

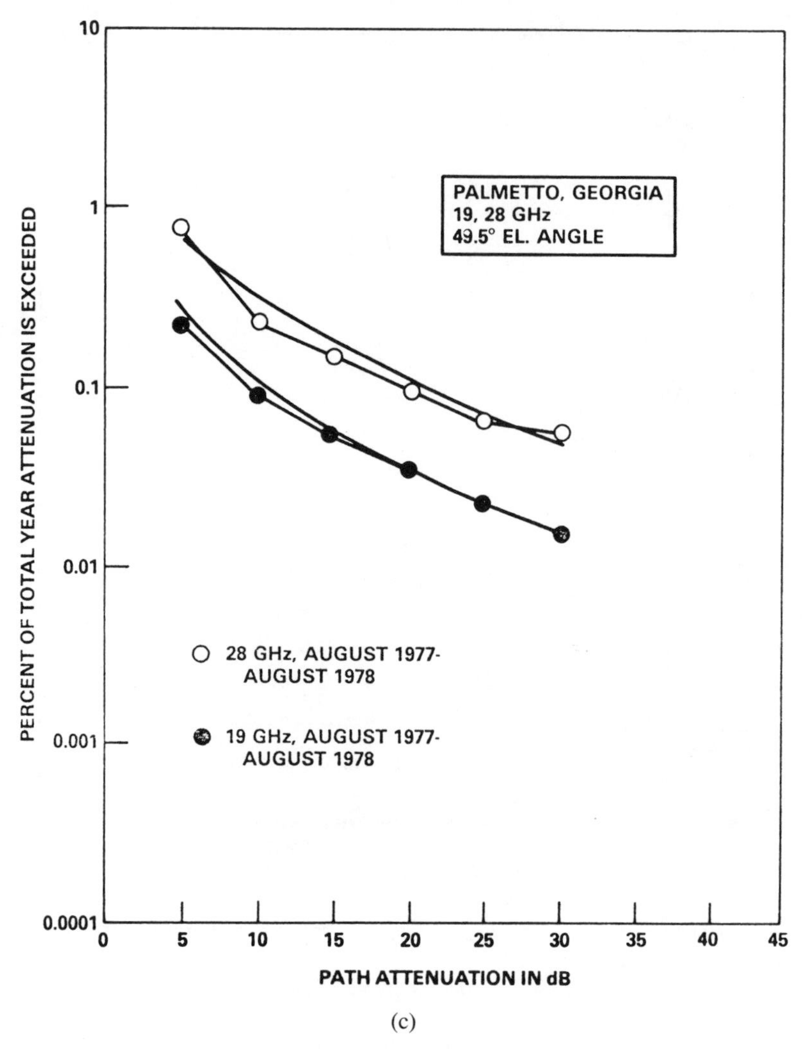

(c)

Figure 5-2. (*Continued*)

and Europe, respectively, as published by Crane [5.16]. The point rain rate distributions for the rain climate regions of the Crane global model are listed in Table 5-1.

The Crane model relates the surface point rain rate R_p to a path averaged rain rate $\overline{R}$ through an effective path average factor, determined empirically from terrestrial measurements of rain rate at path lengths up to 22.5 km. The resulting relationship was modeled by a power law expression

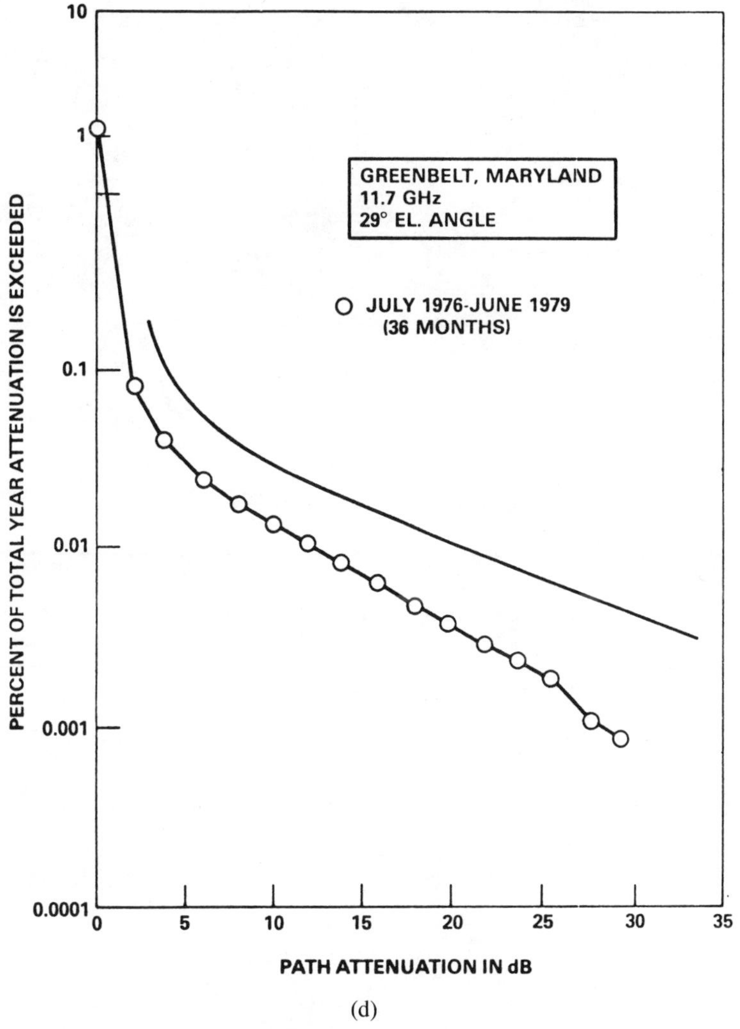

(d)

Figure 5-2. *(Continued)*

$$\bar{R} = \gamma(D)\, R_p^{[1\ +\ \delta(D)]} \tag{5-9}$$

where D is the horizontal path length and γ and δ are determined from a best fit analysis of the terrestrial path data. Relative path profile curves for rain rate as a function of D were determined by numerically differentiating γ and δ, then represented by two exponential functions over the range of D from 0 to 22.5 km. The resultant expression for integrated attenuation for a slant path at an

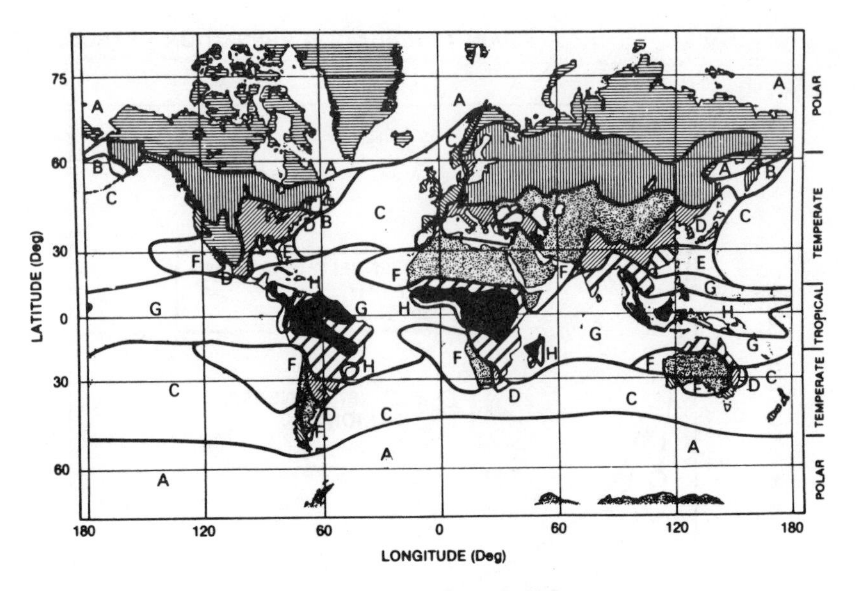

RAIN RATE CLIMATE REGIONS

POLAR:		TEMPERATE:		SUB TROPICAL:		TROPICAL:	
A	Tundra (Dry)	C	Maritime	E	Wet	G	Moderate
B	Taiga (Moderate)	D	Continental	F	Arid	H	Wet

Figure 5-3. Rain climate regions for the Crane global model.

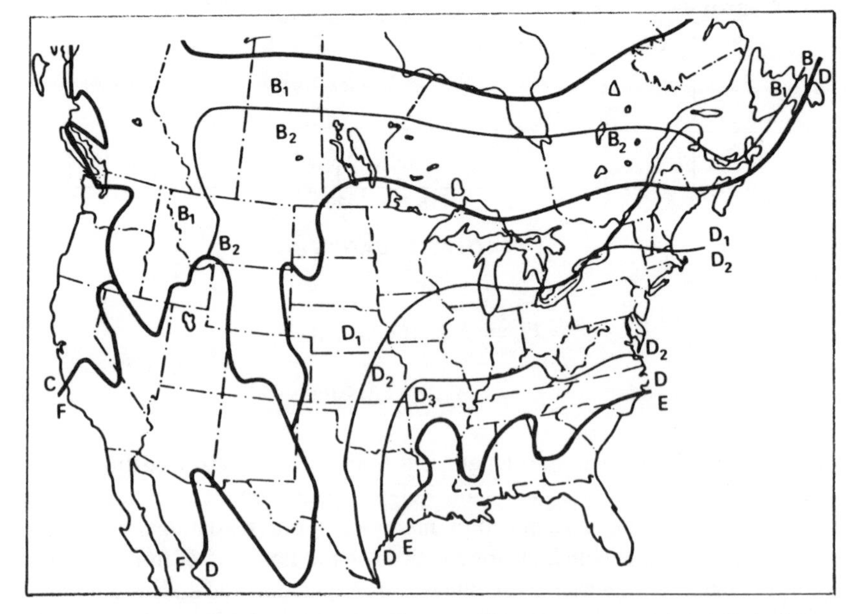

Figure 5-4. Crane global model rain climate regions for the continental United States and southern Canada.

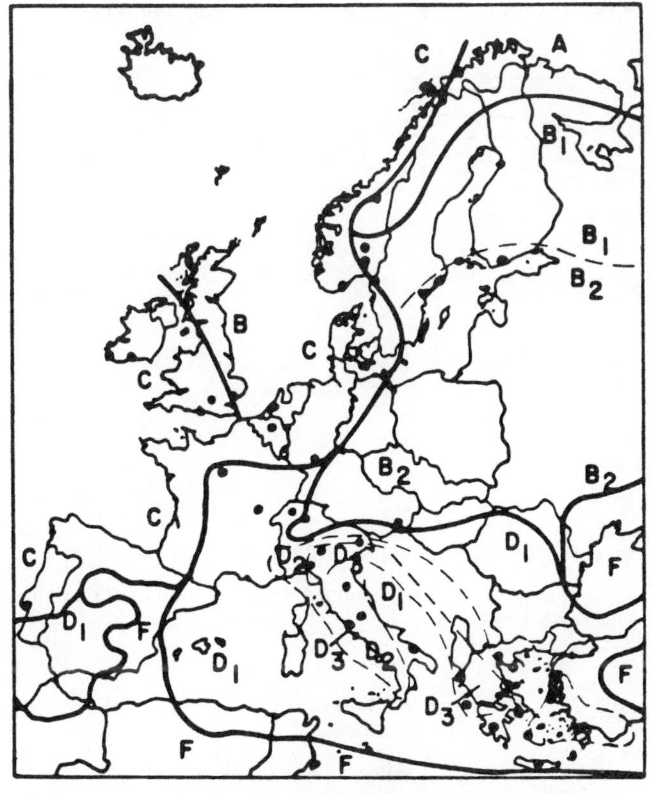

Figure 5-5. Crane global model rain climate regions for Europe.

Table 5-1. Point Rain Rate Distributions for the Rain Climate Regions of the Crane Global Model.

Rain rate exceeded, percent of year	Point rain rate distribution values (mm/hr) per rain climate region									
	A	B	C	D_1	D_2	D_3	E	F	G	H
0.001	28	54	80	90	102	127	164	66	129	251
.002	24	40	62	72	86	107	144	51	109	220
.005	19	26	41	50	64	81	117	34	85	178
.01	15	19	28	37	49	63	98	23	67	147
.02	12	14	18	27	35	48	77	14	51	115
.05	8	9.5	11	16	22	31	52	8.0	33	77
.1	6.5	6.8	72	11	15	22	35	5.5	22	51
.2	4.0	4.8	4.8	7.5	9.5	14	21	3.8	14	31
.5	2.5	2.7	2.8	4.0	5.2	7.0	8.5	2.4	7.0	13
1.0	1.7	1.8	1.9	2.2	3.0	4.0	4.0	1.7	3.7	6.4
2.0	1.1	1.2	1.2	1.3	1.8	2.5	2.0	1.1	1.6	2.8

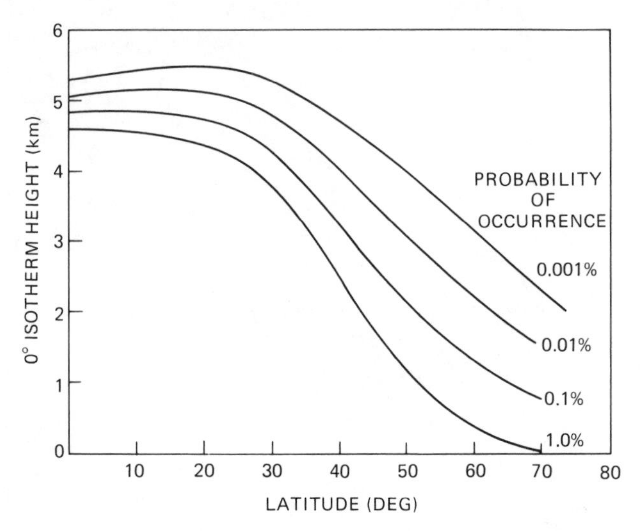

Figure 5-6. 0° isotherm height as a function of ground station location (latitude) for the Crane model.

elevation angle θ, as a function of the point rain rate R, is of the form

$$A(\text{dB}) = \frac{aR_p^b}{\cos\theta}\left[\frac{e^{Ubd}-1}{Ub} - \frac{X^b\, e^{Ybd}}{Yb} + \frac{X^b\, e^{YbD}}{Yb}\right] \tag{5-10}$$

where U, X, and Y are empirical constants that depend on R_p.

The vertical variation of rain rate is accounted for by assuming that R is a constant from the surface to the 0° isotherm height. The 0° isotherm height is a statistically determined variable which is a function of station location (latitude) and the probability of occurrence, as shown in Figure 5-6.

The complete step-by-step procedure for the application of the Crane global model is given in Appendix D.

Upper and lower bounds on the annual attenuation distribution prediction are included in the Crane model. They were determined by combining the statistical variances of each step in the model development.

Figure 5-7(a–f) presents the Crane global model mean prediction and bounds for several locations and frequencies, and compares the model with measured attenuation distributions. All of the predictions show excellent correlation with the measured data, except for the Blacksburg distributions at 28.56 GHz. The Global model develops a prediction curve down to 0.001% and tends to reproduce the tails of the distribution consistently well.

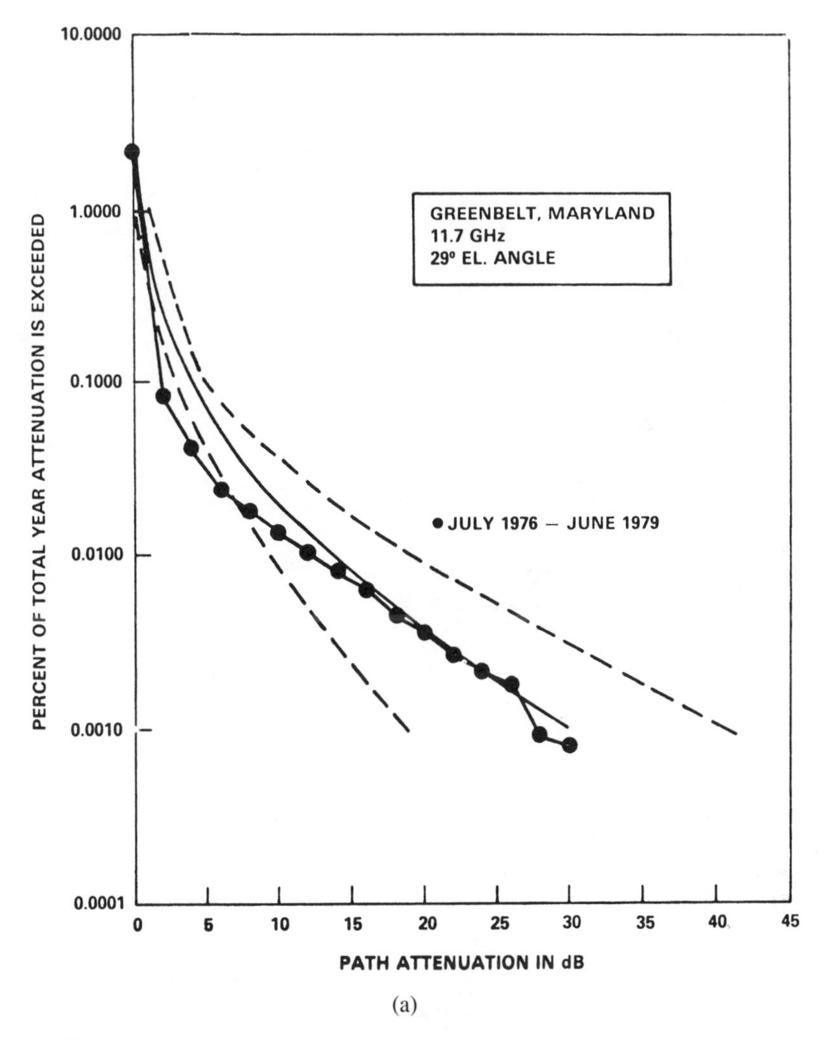

(a)

Figure 5-7. Comparison of the Crane global model with measured data.

5.5. CCIR RAIN ATTENUATION MODELS

The International Radio Consultative Committee, CCIR, at its 15th plenary assembly in Geneva in February 1982, adapted a procedure for the prediction of attenuation caused by rain [5.17]. This event, preceeded by several years of intense deliberations by representatives of the United States, Canada, France, the United Kingdom, the Federal Republic of Germany, and other countries,

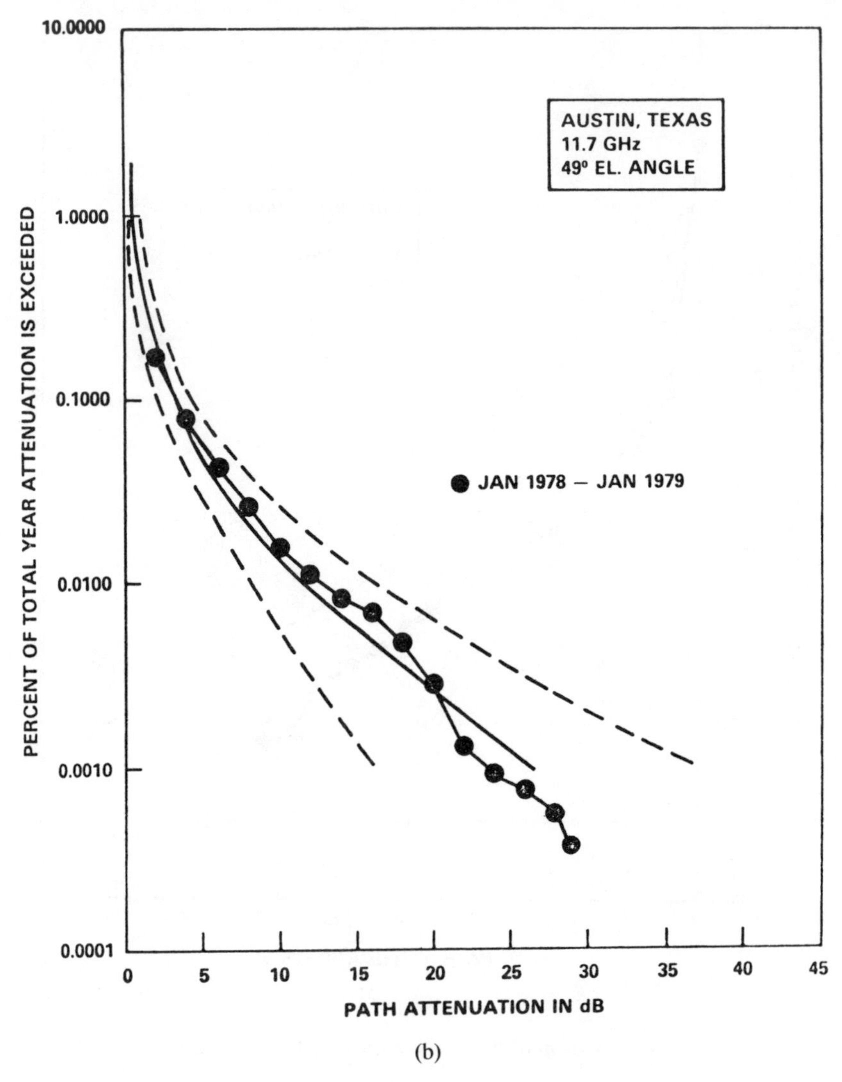

(b)

Figure 5-7. (*Continued*)

marked the first time that a global rain attenuation prediction method had been accepted by an international body.

The procedure provides the basis for rain attenuation calculations required for international planning and coordination meetings, such as the 1983 Regional Administrative Radio Conference for the Broadcasting Satellite Service in Region 2, RARC-83 BSS, and the 1985 World Administrative Radio Conference,

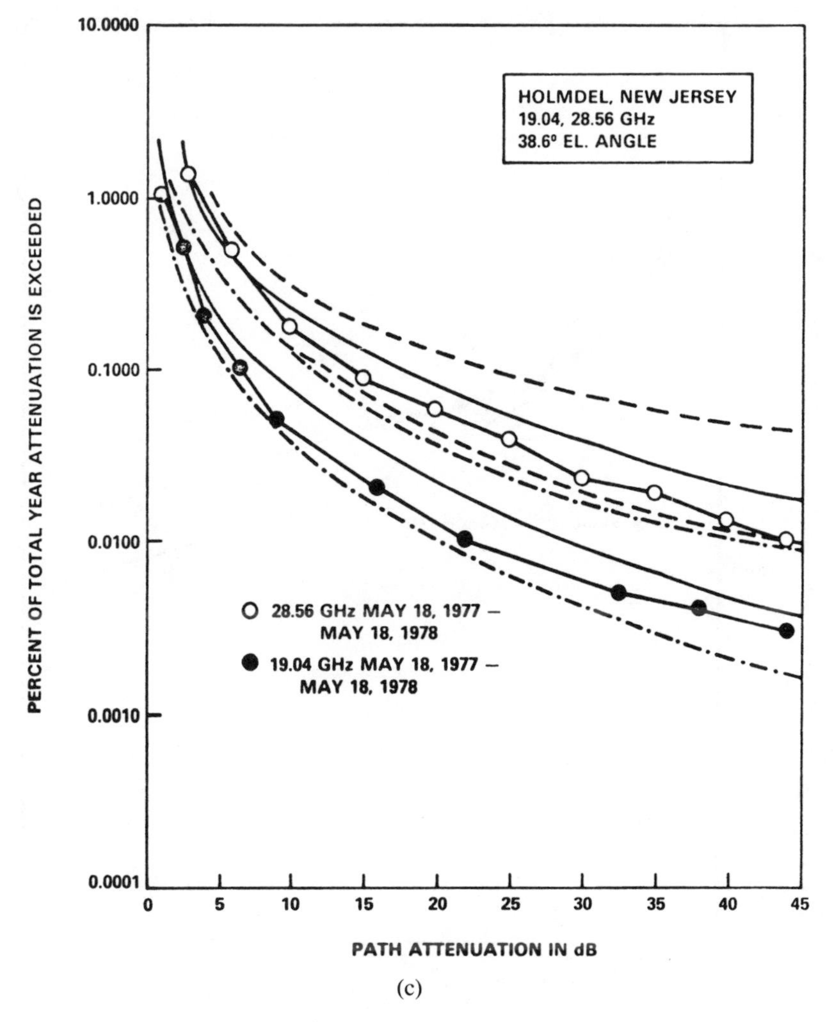

(c)

Figure 5-7. (*Continued*)

WARC-85. The CCIR procedure is not only important in international communications satellite planning, but is also a useful procedure for link design and performance evaluation of domestic systems as well.

The original CCIR procedure was modified in June 1982 at the Conference Preparatory Meeting (CPM) for RARC-83 BSS [5.18]. The modifications, based on recommendations of an Interim Working Party (IWP 5/2) Meeting in May of 1982, were added to improve rain attenuation predictions in tropical climates where the original procedure was found to predict higher values than antici-

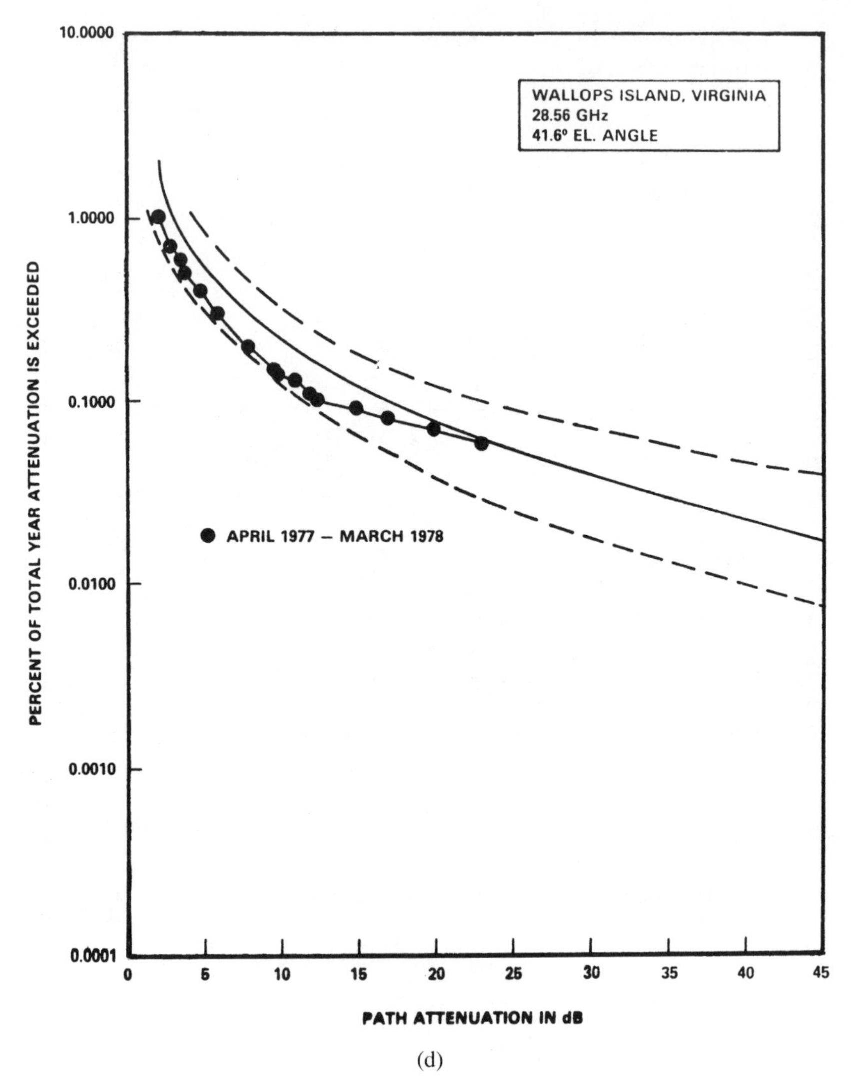

(d)

Figure 5-7. (*Continued*)

pated. The original CCIR procedure consists of two methods, one for maritime climates, defined here as Method I, and one for continental climates and/or for time percentages greater than 0.1%, defined here as Method II. The modifications developed by IWP 5/2 were to Method I, and the resulting procedure is referred to here a Method I′.

The CCIR procedure determines an annual attenuation distribution at a spec-

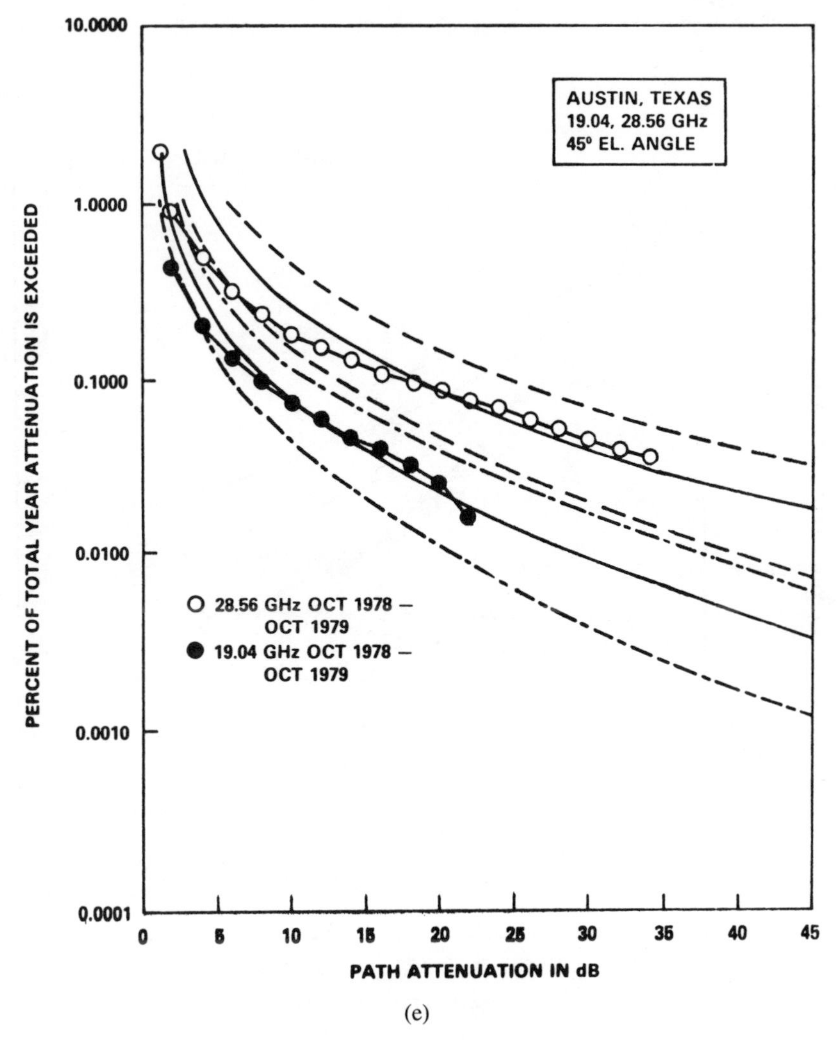

(e)

Figure 5-7. *(Continued)*

ified frequency, elevation angle, polarization and location, from an "average year" rain rate distribution. The input parameters required for the determination of the slant-path attenuation at a given location are:

Frequency, GHz
Elevation angle to the satellite, deg.
Ground station elevation, i.e., height above mean sea level, km
Ground station latitude, deg.

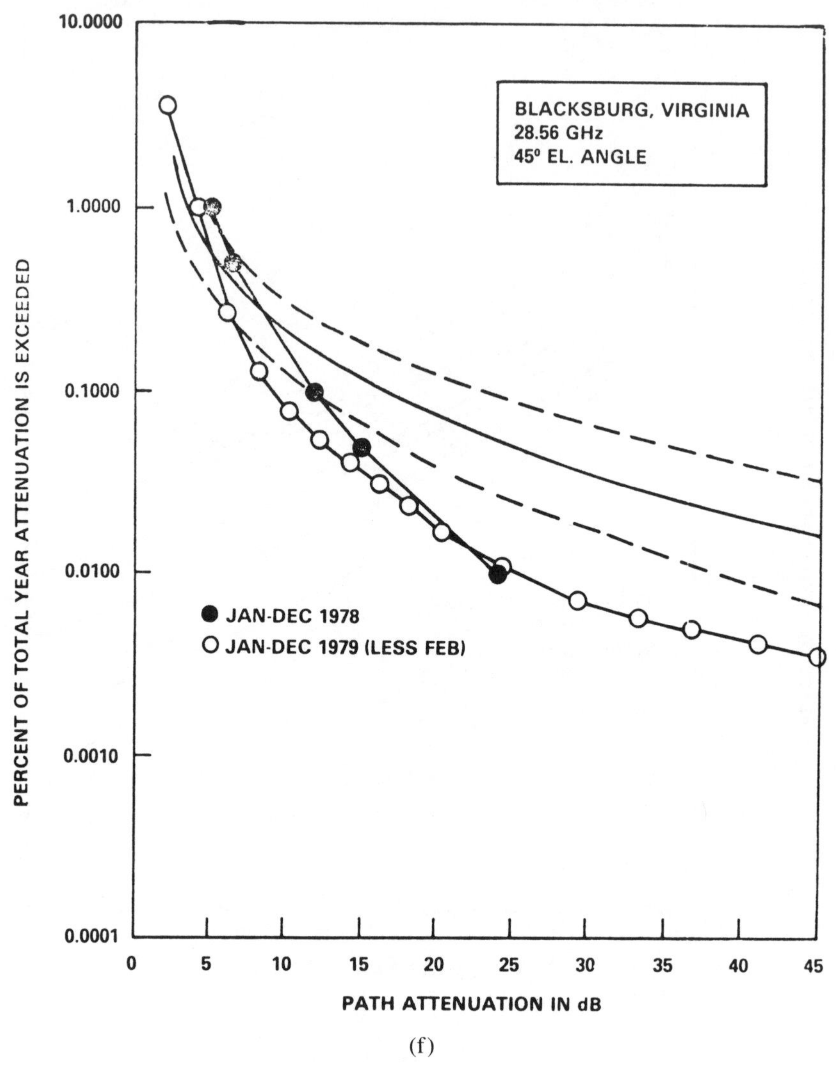

(f)

Figure 5-7. (*Continued*)

The point rainfall rate for the location is required at 0.01% of an average year for Methods I and I′, and at 1%, 0.1%, 0.01%, and 0.001% for Method II. If this rain rate information is not available from local sources, an estimate can be obtained from global maps of climate zones of rainfall rate distributions provided by the CCIR in Report 563-2 [5.19], and reproduced here as Figure 5-8(a–c). A more detailed map of the CCIR climate zones for the continental

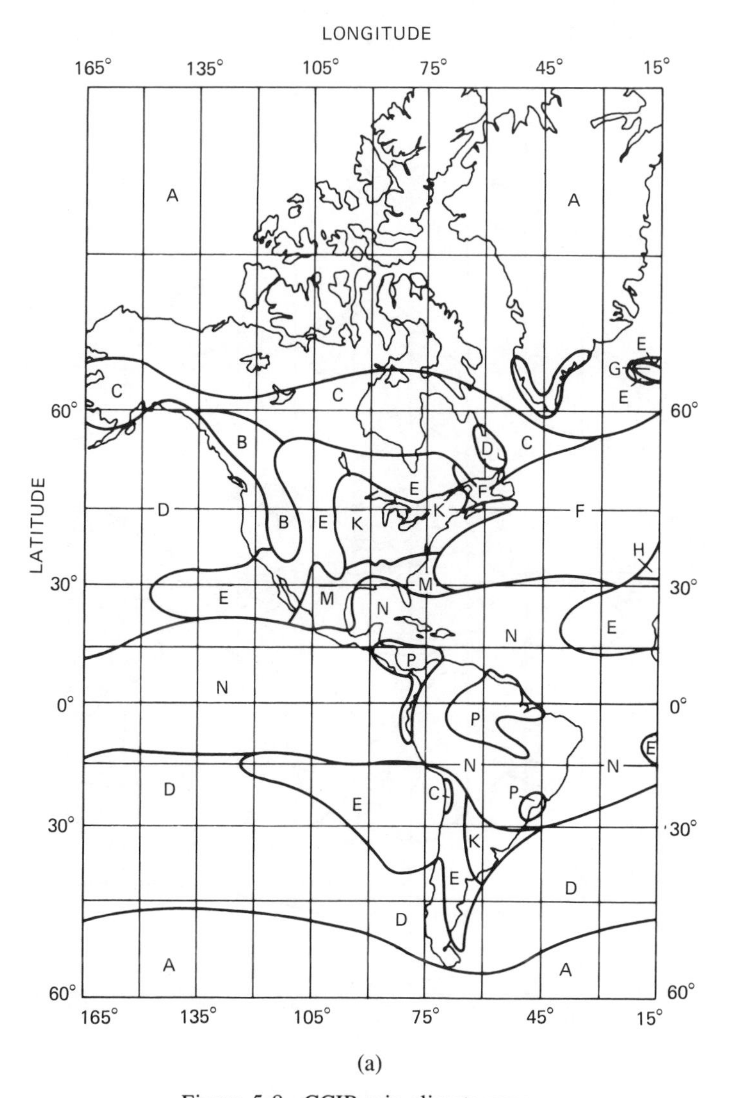

(a)

Figure 5-8. CCIR rain climate zones.

United States and southern Canada is shown in Figure 5-9. The CCIR global rainfall rate map consists of fourteen rain climate zones, *A* through *P*, described by the distributions given in Table 5-2.*

Method I is recommended by the CCIR for maritime climates, such as the

*The climate zones and rain rate distributions of the CCIR model are *not* the same as the Crane global model climate zones and distributions discussed in the preceeding section.

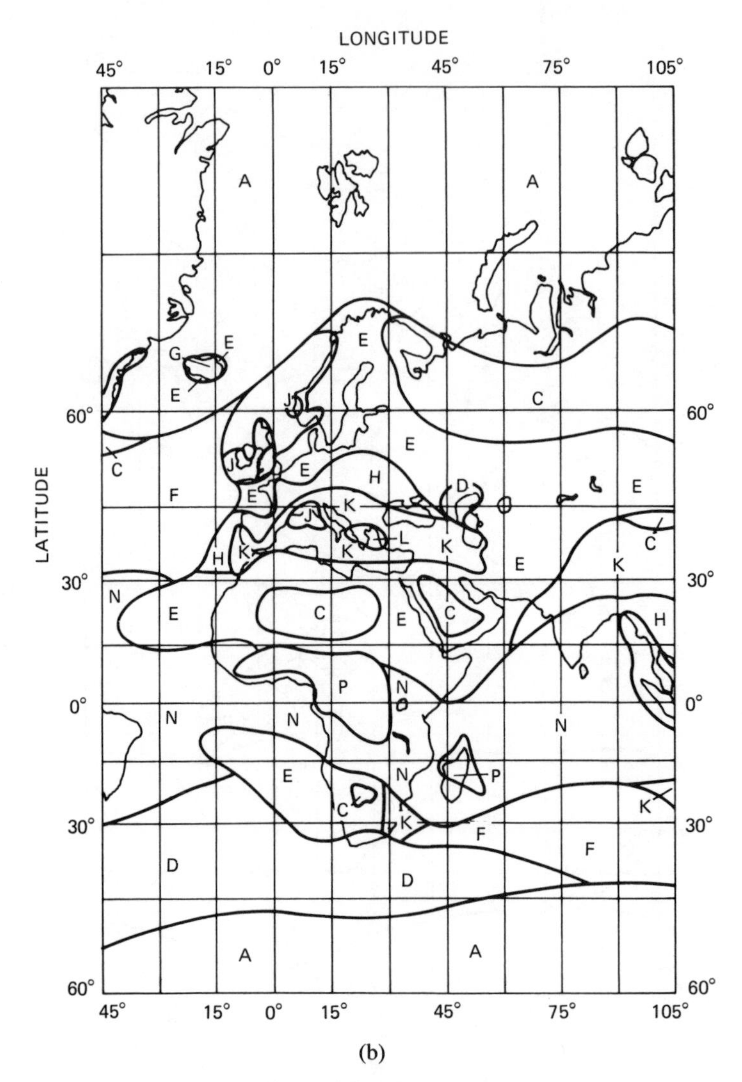

(b)

Figure 5-8. (*Continued*)

coastal areas of the United States for Europe. Method II is recommended for continental climates, the inland areas of the temporate regions of the globe. Method II is also required if rain attenuation predictions for annual time percentages of greater than 0.1% are required, since they are not available from Method I. Method I′ is a refinement of Method I, particularly for tropical regions near the equator. Unfortunately, the CCIR literature offers no guidance on how to select the method for a particular location of interest, or on the

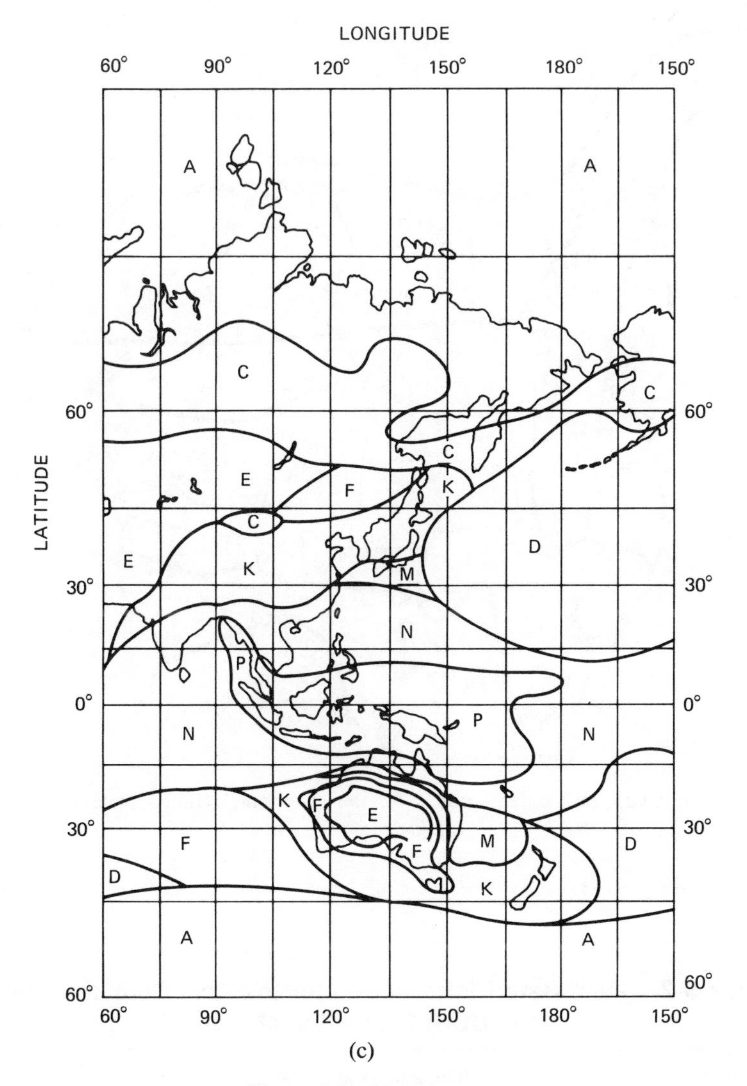

(c)

Figure 5-8. *(Continued)*

relationship of the methods to the various climate zones defined in the CCIR maps. Because of this, it is advisable to calculate the rain attenuation by all three methods, and then to evaluate the prediction from a comparison of the results.

The complete step-by-step procedure for application of the CCIR model is given in Appendix E.

Figure 5-10(a–d) presents comparisons of the three CCIR Methods with mea-

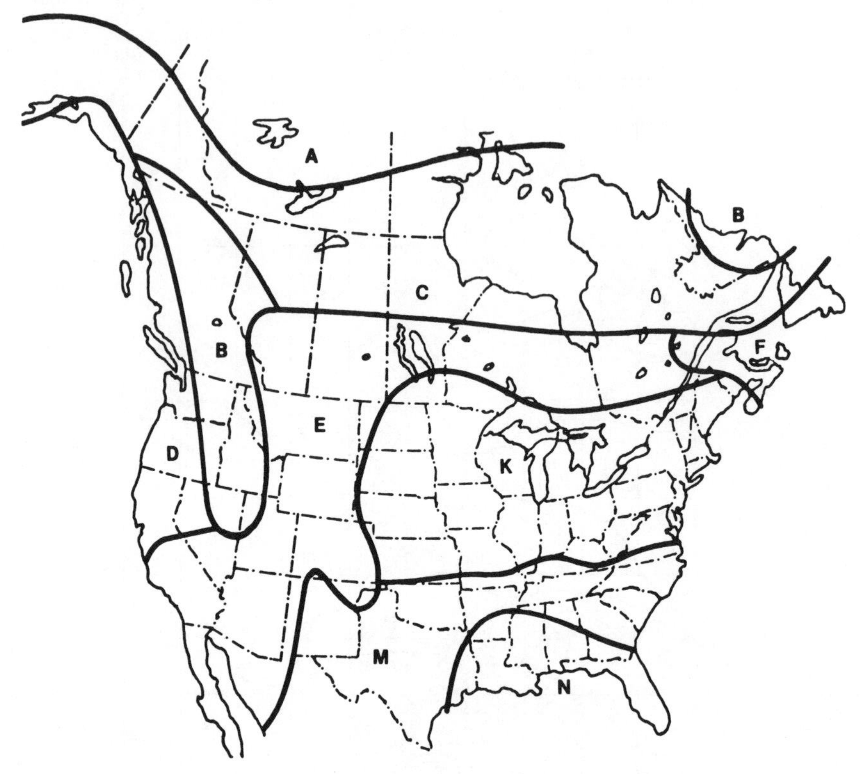

Figure 5-9. CCIR rain climate zones for the continental United States and southern Canada.

Table 5-2. Rain Rate Distributions for the Rain Climate Zones of the CCIR Prediction Model.

Percentage of Time (%)	Rain Rate Distribution Values (mm/h)													
	A	B	C	D	E	F	G	H	J	K	L	M	N	P
1.0	—	1	—	3	1	2	—	—	—	2	—	4	5	12
0.3	1	2	3	5	3	4	7	4	13	6	7	11	15	34
0.1	2	3	5	8	6	8	12	10	20	12	15	22	35	65
0.03	5	6	9	13	12	15	20	18	28	23	33	40	65	105
0.01	8	12	15	19	22	28	30	32	35	42	60	63	95	145
0.003	14	21	26	29	41	54	45	55	45	70	105	95	140	200
0.001	22	32	42	42	70	78	65	83	55	100	150	120	180	250

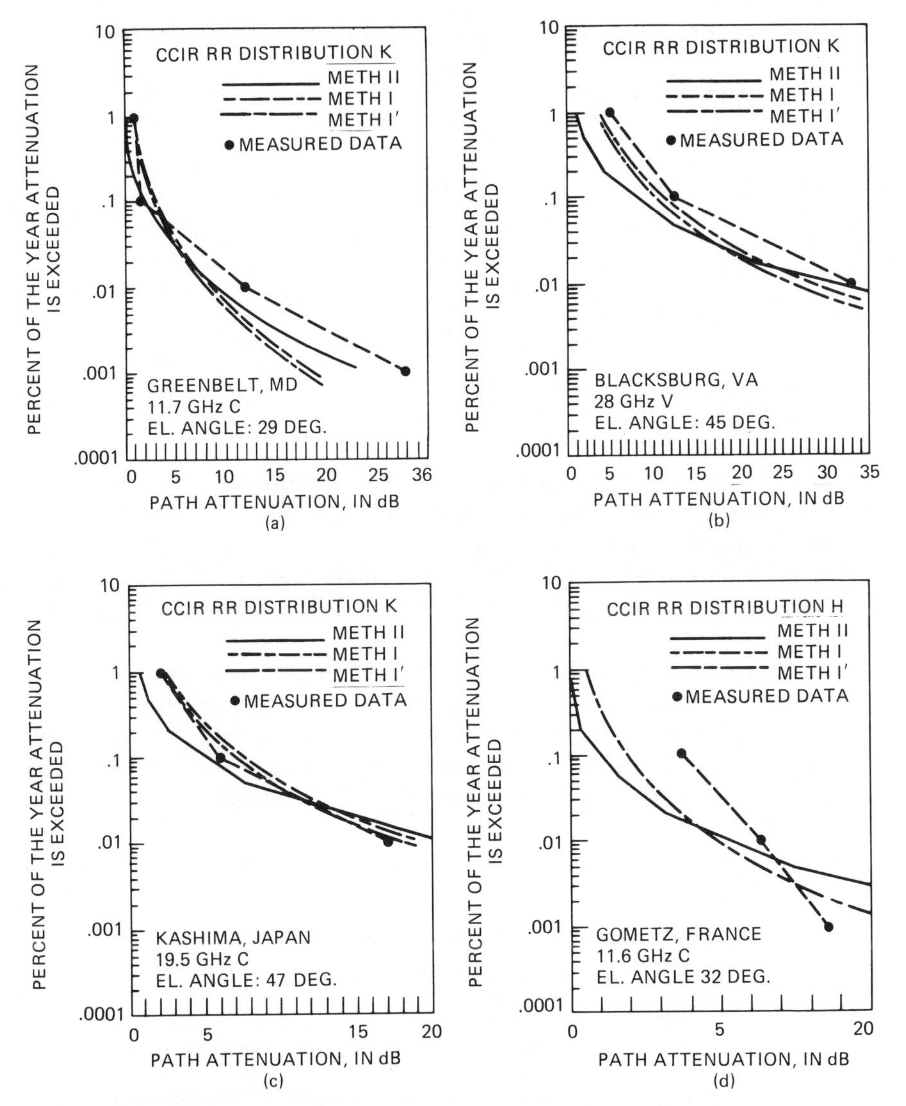

Figure 5-10. Comparison of the CCIR procedure with measured data.

sured distributions in the United States, Europe, and Japan, at frequencies from 11.6 to 28 GHz. The results show reasonably good agreement, except for the measurements in Europe.

The prediction curves of Figure 5-10 also point out several of the general characteristics of the three CCIR methods. Method II tends to consistently predict lower attenuation values than Methods I or I′ for percentages of the year

Table 5-3. Comparison of CCIR Prediction Methods for Global Locations.

		Method II		Method I		Method I′	
Time Percentage	Number of Data Sets Used	Mean Difference (%)	Standard Deviation (%)	Mean Difference (%)	Standard Deviation (%)	Mean Difference (%)	Standard Deviation (%)
0.001%	24	25	41	14	60	2	41
0.01%	74	10	41	10	41	−3	30
0.1%	66	0	83	37	71	17	50
1.0%	39	−43	66	77	160	39	113

*48 sets with measured rain rate data.
33 sets with CCIR rain rates.

about 0.01% or greater. For percentages below about 0.01% that is, for the high attenuation values or tails of the distributions, Method II tends to predict higher attenuation values than Methods I or I′. This result is observed for all combinations of elevation angle, frequency, latitude, and climate zone, except for very low elevation angle/very high latitude locations, where the trend is reversed.

An extensive analysis of the CCIR prediction procedure was performed with 102 data sets from 44 global locations [5–20]. The data sets included frequencies from 4 GHz to above 30 GHz, with 43 from the United States, 27 from Europe, 22 from Japan, and 10 from other areas. Most of the measurements were from satellite beacons on the CTS, COMSTAR, SIRIO, ETS-II, CS, or OTS satellites.

Table 5-3 shows a comparison of the three CCIR methods with the full global

Table 5-4. Comparison of CCIR Prediction Methods for United States Locations.

	Prediction Using CCIR Rain Rate Distribution			Prediction Using Measured Rain Rate Distribution		
	Method II	Method I	Method I′	Method II	Method I	Method I′
Mean error (%)	−12	39	33	4	78	77
Standard deviation (%)	58	80	75	83	123	126
Number of *Over* Predictions	13	30	28	17	35	32
Number of *Under* Predictions	26	9	11	22	4	7

data set, provided by IWP 5/2, at four time percentages. Method I′ gives the best overall agreement, except at 0.1%, where Method II is better.

A different conclusion is reached for comparisons with the United States data set. Table 5-4 shows a comparison of the CCIR Methods for 39 U.S. locations which had coincident rain rate measurements along with the rain attenuation measurements. Method II is shown to be a better predictor by a factor of two or more when CCIR rain rate distributions are used, and provides only a 4% error when measured rain rate distributions are used. Methods I and I′ tend to overpredict most of the time, while Method II is more evenly distributed between over- and underprediction. The results for the U.S. locations are similar when the measurements are analyzed by frequency, percent of year, location, or elevation angle.

5.6. SUMMARY

A review of several of the more promising rain attenuation prediction models has been presented and representative predictions compared with measured data in the frequency bands important for satellite communications. While all of the models have some advantages, the Crane global model provides the better overall prediction method for U.S. locations, based on its ease of applicability, the inclusion of error bounds, and the good correlation with measured attenuation, particularly in the tails of the distribution. The CCIR Method II procedure also gives good results for the U.S., particularly when measured rain rate distributions are used. The CCIR Method I′ procedure gives some improvement over the Method I procedure, and provides better predictions when global locations are considered.

REFERENCES

5.1. CCIR, Report 563, "Radio-Meteorological Data," Geneva, 1972.
5.2. CCIR, Report 563-1, "Radio-Meteorological Data," Doc. 5/1022, XIVth Plenary Assembly, Kyoto, Japan, 1978.
5.3. Rice, P. L. and Holmberg, N. R., "Cumulative Time Statistics of Surface Point-Rainfall Rates," *IEEE Trans. on Communications*, Vol. COM-21, No. 10, October 1973.
5.4. Dutton, E. J. and Dougherty, H. T., "Modeling the Effects of Cloud and Rain Upon Satellite-to-Ground System Performance," OT Report 73-5, Office of Telecommunications, Boulder, CO, March 1973.
5.5. Dutton, E. J., "Earth Space Attenuation Prediction Procedure at 4 to 16 GHz," OT Report 77-123, Office of Telecommunications, Boulder, CO, May 1977.
5.6. Grantham, D. D. and Kantor, A. J., "Distribution of Radar Echos over the United States," Air Force Surveys in Geophysics, No. 1191, AFCRL-67-0232, March 1977.
5.7. Dutton, E. J., "A Meteorological Model for use in the Study of Rainfall Effects on Atmosphere Radio Telecommunications," Report OT/TREP-24, Office of Telecommunications, Boulder, CO, December 1971.

5.8. Crane, R. K., ''Microwave Scattering Parameters for New England Rain,'' MIT Lincoln Labs., Tech Report 426, October 1966.
5.9. Dutton, E. J., ''Earth Space Attenuation Predictions for Geostationary Satellite Links in the U.S.A.,'' NTIA Report 78-10, U.S. Dept. of Commerce, Boulder, CO, October 1978.
5.10. Dutton, E. J., Kobayashi, H. K., and Dougherty, H. T., ''An Improved Model for Earth–Space Microwave Attenuation Distribution Prediction,'' *Radio Science*, Vol. 17, No. 6, pp. 1360–1370, Nov.-Dec. 1982.
5.11. ''Telecommunications Analysis Services—Reference Guide,'' U.S. Dept. of Commerce, NTIA, Institute for Telecommunication Sciences, Spectrum Utilization Division, Boulder, CO, June 2, 1981.
5.12. Lin, S. H., ''Nationwide Long-Term Rain Rate Statistics and Empirical Calculation of 11 GHz Microwave Rain Attenuation,'' *The Bell System Technical Journal*, Vol. 56, No. 9, November 1977.
5.13. Lin, S. H., ''Empirical Rain Attenuation Model for Earth–Satellite Paths,'' *IEEE Trans. on Communications*, Vol. COM-27, No. 5, May 1979.
5.14. Crane, R. K., ''A Global Model for Rain Attenuation Prediction,'' *EASCON' 78 Record*, IEEE Pub. 78CH 1354-4 AES, Arlington, VA, Sept. 1978.
5.15. Crane, R. K., ''Prediction of Attenuation by Rain,'' *IEEE Trans. on Communications*, Vol. COM-28, pp. 1717-1733, Sept. 1980.
5.16. Crane, R. K., ''A Two-Component Rain Model for the Prediction of Attenuation Statistics,'' *Radio Science*, Vol. 17, No. 6, pp. 1371–1388, Nov.-Dec. 1982.
5.17. CCIR, Report 564-2, ''Propagation Data Required for Space Telecommunications Sytems,'' in Volume V, *Propagation in Non-Ionized Media*, Recommendations and Reports of the CCIR—1982, International Telecommications Union, Geneva, pp. 331–373, 1982.
5.18. CCIR, Report of the CCIR Conference Preparatory Meeting, ''Technical Bases for the Regional Administrative Radio Conference—1983 for the Plannning of the Broadcasting Satellite Service in Region 2,'' Geneva, 28 June–9 July 1982.
5.19. CCIR, Report 563-2, ''Radiometeorological Data,'' in Volume V, *Propagation in Non-Ionized Media*, Recommendations and Reports of the CCIR—1982, International Telecommunications Union, Geneva, pp. 96–123, 1982.
5.20. Ippolito, L. J., ''Application and Evaluation of the CCIR Rain Attenuation Prediction Model for Earth-Space Paths,'' National Radio Science Meeting, Boulder, CO, Jan. 5, 1983.

CHAPTER 6
DEPOLARIZATION ON SATELLITE PATHS

The earth's atmosphere can produce changes in the polarization sense of transmitted radiowaves. This effect is referred to as *depolarization*, or cross-polarization in the case of two independent waves which interfere with each other because of a depolarizing condition present in the transmission path.

A knowledge of the depolarizing characteristics of the earth's atmosphere is particularly important in the design of frequency reuse communications systems employing dual independent orthogonal polarized channels in the same frequency band to increase channel capacity. Frequency reuse techniques, which employ either linear or circular polarized transmissions, can be impared by the propagation path through a transferral of energy from one polarization state to the other orthogonal state, resulting in interference between the two channels.

Depolarization can be induced on a satellite link from two basic conditions on the path: (1) *hydrometeors*, primarily rain and ice crystals, and (2) *multipath propagation*, either tropospheric or ionospheric, or from surface and terrain irregularities.* Hydrometeor depolarization is generally a problem at frequencies above 3 GHz, while multipath depolarization occurs in the frequency bands below 3 GHz.

Radiowave depolarization is characterized by the presence of an anisotropic propagation medium which produces a different attenuation and phase shift on radiowaves with different polarizations. The wave will have its polarization state altered such that power is transferred (or coupled) from the desired polarization state to the undesired orthogonal polarization state, resulting in interference or crosstalk between the two orthogonally polarized channels.

Figure 6-1 shows a pictoral representation of the depolarization effect in terms of the *E*-field (electric field) vectors in a linearly polarized transmission link. The vectors E_1 and E_2 are the transmitted waves which are polarized 90° apart (orthogonal) to provide two independent signals at transmission. The transmit-

* The *Faraday Effect*, which is caused by ionospheric variations of the earth's magnetic field, produces a polarization *rotation* only, on linearly polarized VHF links. The rotation can usually be compensated for.

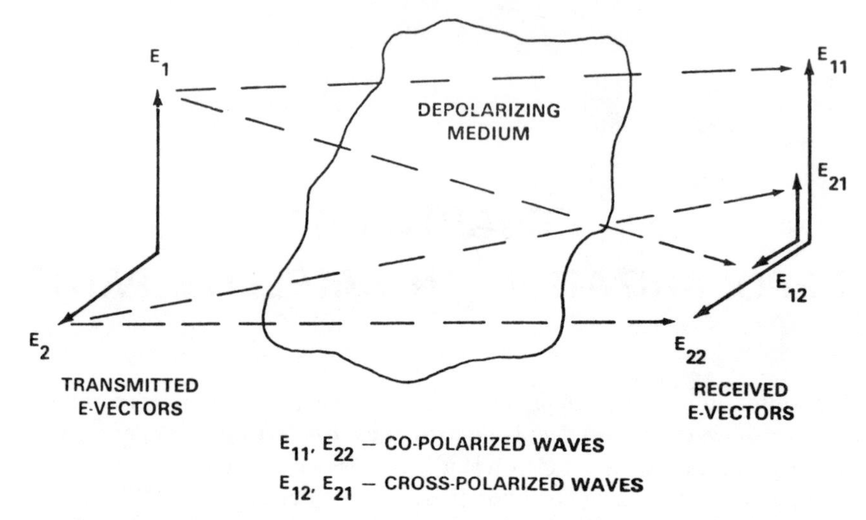

Figure 6-1. Depolarization on a linearly polarized radiowave transmission link.

ted waves will be depolarized by the medium into several components, as shown in Figure 6-1.

The *cross-polarization discrimination XPD* is defined for the linearly polarized waves as

$$XPD = 20 \log \frac{|E_{11}|}{|E_{12}|} \tag{6-1}$$

where E_{11} is the received electric field in the copolarized (desired) direction and E_{12} is the electric field converted to the orthogonal cross-polarized (undesired) direction. A closely related parameter is the *isolation I*, which compares the copolarized recieved power with the cross-polarized power received in the same polarization state, i.e.,

$$I = 20 \log \frac{|E_{11}|}{|E_{21}|} \tag{6-2}$$

Isolation takes into account the performance of the receiver antenna, feed and other components, as well as the propagating medium. When the received system polarization performance is close to ideal, then *XPD* and *I* are nearly identical, and only the propagating medium contributes depolarizing effects to system performance.

Unfortunately, the terms used to describe depolarization parameters are not always applied as universally accepted definitions. Some authors use the reciprical of the definitions above for *XPD* and *I*, or refer to *XPD* as *I*, or use other terms such as cross-polarization distortion, depolarization ratio, or cross-talk discrimination, among several others. Care should be taken to determine just how the term being used is defined, preferably in terms of the vector components as described above.

The remaining sections of this chapter discuss the characteristics of rain depolarization, ice depolarization, and multipath depolarization in more detail. Methods of calculating the expected depolarization parameters are given, and measured data on earth–space links are reviewed and summarized.

6.1. DEPOLARIZATION CAUSED BY RAIN

Rain induced depolarization is produced from a differential attenuation and phase shift caused by nonspherical raindrops. As the size of raindrops increase, their shape tends to change from spherical (the preferred shape because of surface tension forces) to oblate spheroids with an increasingly pronounced flat or concave base caused by aerodynamic forces acting upward on the drops [6.1]. Furthermore, raindrops may also be inclined to the horizontal (canted) because of vertical wind gradients. Therefore the depolarization characteristics of a linearly polarized radiowave will depend significantly on the transmitted polarization angle.

The classical model for a falling raindrop is an oblate spheroid with its major axis canted to the horizontal and with major and minor axes related to the radius of a sphere of equal volume [6.2, 6.3]. Consider a volume of canted oblate spheroidal raindrops of extent L in the direction of propagation of two orthogonally polarized linear waves, E_1 (vertical) and E_2 (horizontal), as shown in Figure 6-2. The drops are randomly canted at an angle ϕ to the horizontal, with minor and major axes a and b respectively, in the directions I and II. The transmitted waves can be resolved into components in the I and II directions by

$$E_{TI} = E_1 \cos \phi - E_2 \sin \phi$$

$$E_{TII} = E_1 \sin \phi + E_2 \cos \phi \tag{6-3}$$

The transmission characteristics of the volume of oblate spheroids are specified by transmission coefficients in the I and II directions of the form

$$T_I = e^{-(A_I - i\,\Phi_I)L}$$

$$T_{II} = e^{-(A_{II} - i\,\Phi_{II})L} \tag{6-4}$$

where A_I and A_{II} are the attenuation coefficients and Φ_I and Φ_{II} are the phase shift coefficients in the minor and major axis directions, respectively.

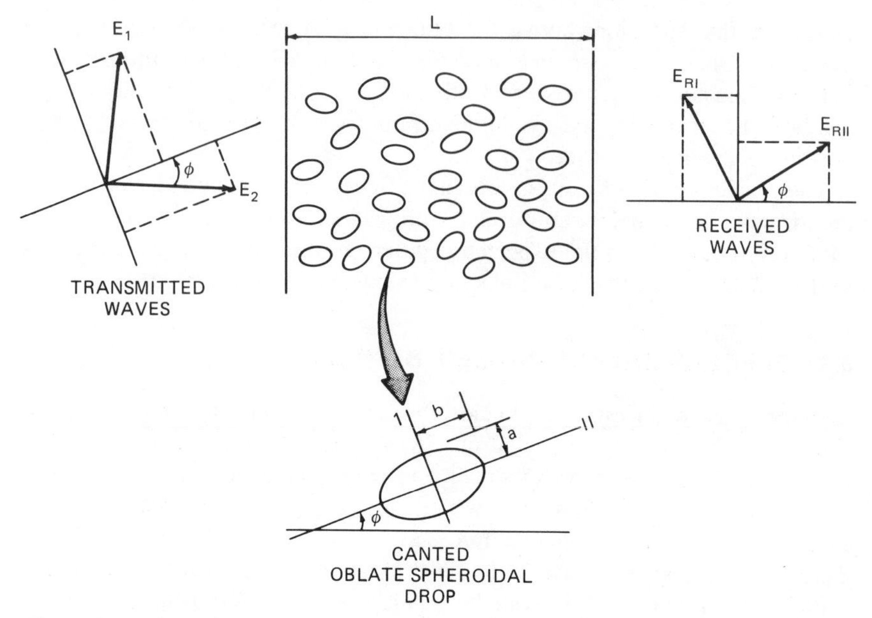

Figure 6-2. Classical model for depolarization by canted oblate spheroidal raindrops.

The received waves in the I and II directions are then found from

$$E_{R\mathrm{I}} = T_{\mathrm{I}}\, E_{T\mathrm{I}}$$

$$E_{R\mathrm{II}} = T_{\mathrm{II}}\, E_{T\mathrm{II}} \tag{6-5}$$

Combining Equations (6-3), (6-4), and (6-5), the resulting received waves in the transmission directions 1 (vertical) and 2 (horizontal) are

$$E_{R1} = a_{11}\, E_1 + a_{21}\, E_2$$

$$E_{R2} = a_{12}\, E_1 + a_{22}\, E_2 \tag{6-6}$$

where a_{11}, a_{12}, a_{21}, and a_{22} are polarization coefficients defined by

$$a_{11} = T_{\mathrm{I}} \cos^2 \phi + T_{\mathrm{II}} \sin^2 \phi$$

$$a_{22} = T_{\mathrm{I}} \sin^2 \phi + T_{\mathrm{II}} \cos^2 \phi$$

$$a_{12} = a_{21} = \left(\frac{T_{\mathrm{II}} - T_{\mathrm{I}}}{2}\right) \sin 2\phi \tag{6-7}$$

The *XPD*'s for *vertical* and *horizontal* transmissions are then given by

$$XPD_v = 20 \log \frac{|a_{11}|}{|a_{12}|}$$

$$= 20 \log \frac{1 + (T_{II}/T_I) \tan^2 \phi}{[(T_{II}/T_I) - 1] \tan \phi} \tag{6-8}$$

$$XPD_H = 20 \log \frac{|a_{22}|}{|a_{21}|}$$

$$= 20 \log \frac{(T_{II}/T_I) + \tan^2 \phi}{[(T_{II}/T_I) - 1] \tan \phi} \tag{6-9}$$

Note that both a_{11} and a_{22} are independent of the sign of ϕ, and also since $a_{12} = a_{21}$, the cross-polarized components resulting from positive and negative canting angles will cancel each other out.

The *XPD* for *circular* transmitted polarization is developed in a similar manner, and is expressed in terms of the polarization coefficients as follows [6.3]:

$$XPD_c = 20 \log \left[\left| \frac{|a_{11}|}{|a_{12}|} \right|_{\phi = 45°} \overline{|e^{i2\phi}|} \right]$$

$$= 20 \log \left[\frac{T_{II} + T_I}{T_{II} - T_I} \overline{|e^{i2\phi}|} \right] \tag{6-10}$$

where $\overline{e^{i2\phi}}$ is the mean value of $e^{i2\phi}$ taken over the canting angle distribution. Equation (6-10) applies to either right hand or left hand circular polarization.

Measurements on earth–space paths using satellite beacons have shown that the average canting angle tends to be very close to 0° (horizontal) for the majority of nonspherical raindrops [6.4]. Under this condition, the *XPD* for circular polarization is identical to the *XPD* for linear horizontal or vertical polarization oriented at 45° from the horizontal.

The complete solution for the *XPD* requires a determination of the transmission coefficients T_I and T_{II}. This problem was examined by Oguchi [6.2] and solved by empolying a spherical expansion solution for a plane wave incident on an oblate spheroid, analogous to the Mie solution for a spherical drop in the rain attenuation case. An equivolume spherical drop is assumed, which approximates the experimentally observed drop shape by a simple function of the major and minor axes.

Figure 6-3 presents the differential attenuation ($A_{II} - A_I$) in dB/km and the differential phase shift ($\Phi_{II} - \Phi_I$) in degrees/km, as a function of frequency for a Laws and Parsons drop size distribution at 20°C, with the direction of prop-

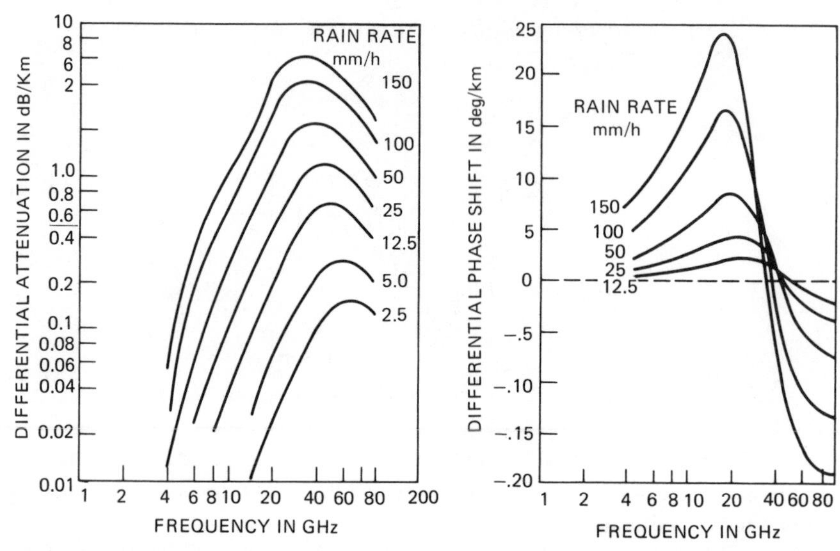

Figure 6-3. Differential attenuation and differential phase shift as a function of frequency and rain rate.

agation at 90° to the axis of symmetry of the raindrop [6.3]. The differential attenuation at 12 GHz, for example, reaches a value of 1.7 dB at 150 mm/h, while at 30 GHz is 6 dB for the same rain rate.

The *XPD*'s for linear horizontal, linear vertical, and circular transmitted waves are shown on Figure 6-4 as a function of copolarized rain attenuation for five frequencies from 4 to 30 GHz, as calculated by Chu for a 100 mm/h rain rate and a 25° canting angle [6.3]. For a given copolarized attenuation, the *XPD* decreases (degrades) with decreasing frequency. Note that 4 and 6 GHz communications systems seldom experience copolarized rain attenuations of more then a few dB, so the *XPD* degradation is not generally a major problem in system performance. Copolarized attenuations can exceed 10–20 dB for 11, 18, and 30 GHz systems, however, so *XPD* degradations of 10–15 dB and greater can occur.

The *XPD* for vertical polarized transmissions is generally higher (better) than for horizontal polarized transmission at the same attenuation level, except at 4 GHz, where the differential phase shift dominates the depolarization effect. Circularly polarized waves produce *XPD*'s about 10 dB lower (worse) than a horizontally polarized wave at the same attenuation level.

6.1.1. Rain Depolarization Prediction

When measurements of depolarization observed on a radiowave path were compared with rain attenuation measurements coincidently observed on the same

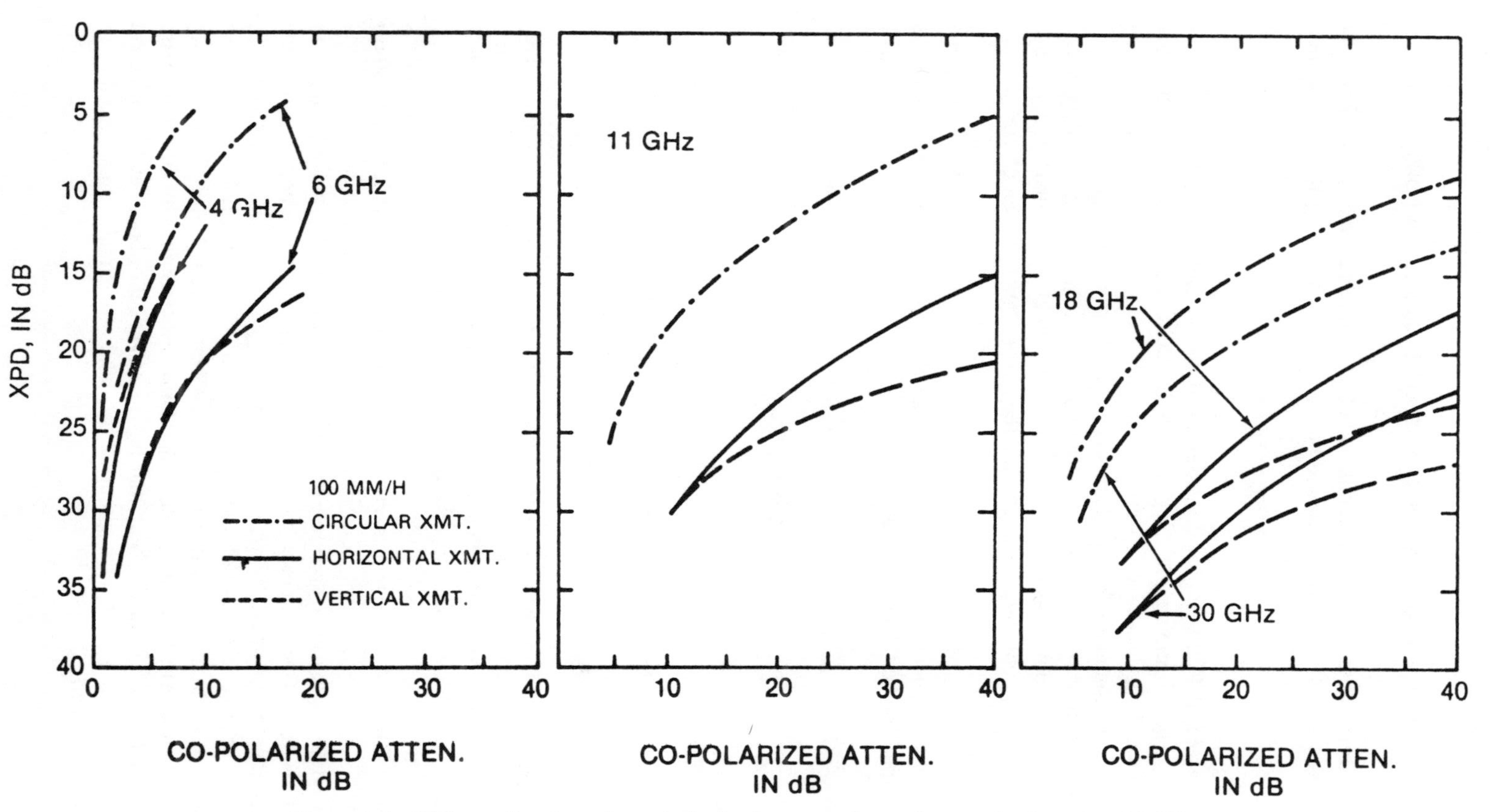

Figure 6-4. *XPD* as a function of copolarized rain attenuation at frequencies from 4 to 30 GHz.

path, it was noted that the relationship between the measured statistics of *XPD* and copolarized attenuation A could be well approximated by the relationship

$$XPD = U - V \log A \tag{6-11}$$

where U and V are empirically determined coefficients which depend on frequency, polarization angle, elevation angle, canting angle, and other link parameters. This discovery is similar to the aR^b relationship observed between rain attenuation and rain rate discussed in the previous chapter.

A theoretical basis for the relationship between rain depolarization and attenuation given above was developed by Nowland et al. from small argument approximations applied to the scattering theory of Oguchi for an oblate spheroid raindrop [6.5].

For most practical applications semi-empirical relations can be used for the U and V coefficients which are valid from 8 to 35 GHz. The coefficients, as specified by the CCIR [6.6], are

$$U = 30 \log f - 10 \log (0.5 - 0.4697 \cos 4\tau) - 40 \log (\cos \theta) \tag{6-12}$$

and

$$V = \begin{cases} 20, & \text{for } 8 < f \leq 15 \text{ GHz} \\ 23, & \text{for } 15 < f \leq 35 \text{ GHz} \end{cases} \tag{6-13}$$

where f is the frequency, in GHz, τ is the tilt angle, in degrees, of the polarization with respect to the horizontal (for circular polarization, $\tau = 45°$), and θ is the elevation angle of the path, in degrees. The elevation angle must be less then or equal to 60° for this relationship to be valid.

Figure 6-5(a,b) presents the *XPD* as calculated from the CCIR coefficients for 12 GHz and 20 GHz, respectively, for copolarized attenuations up to 30 dB and for elevation angles from 10 to 60 degrees. The dependence of *XPD* on elevation angle can be seen to vary by about 12 dB over the range from 10° to 60°, with best performance occuring at the higher elevation angles, where the path length is shortest.

The relative advantage of linear polarization over circular polarization is also apparent in the figures, with the difference being about 15 dB. This improvement factor of linear over circular polarization is determined by the second term in Equation (6-12) for U. Note that for linear vetical or horizontal polarization ($\tau = 90°$ or $0°$) the term is about 15 dB, while for circular polarization ($\tau = 45°$) it is nearly 0 (0.13 dB).

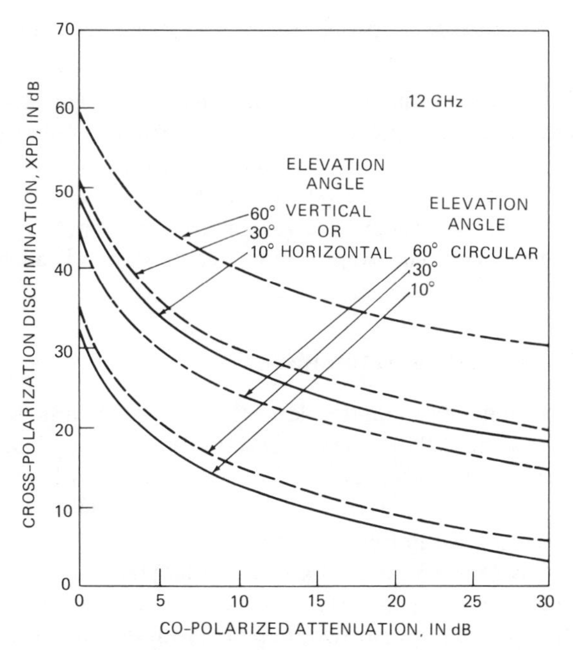

Figure 6-5(a). *XPD* as a function of copolarized attenuation and elevation angle from the CCIR *U* and *V* coefficients, at 12 GHz.

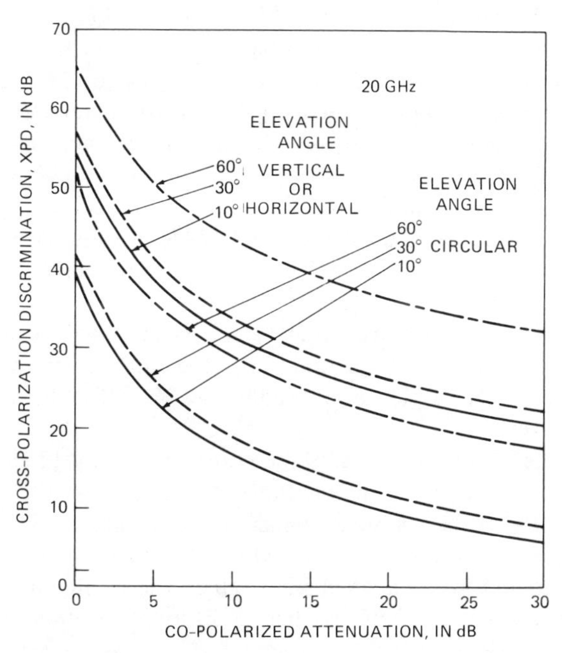

Figure 6-5(b). *XPD* as a function of copolarized attenuation and elevation angle from the CCIR *U* and *V* coefficientss, at 20 GHz.

The CCIR coefficients can be used with Equation (6-11) to obtain *XPD* statistics for linear or circular polarized radiowaves using copolarized rain attenuation statistics developed from the prediction methods described in the previous chapter. It should be noted, however, that the effects of ice depolarization are *not* accounted for in the above analysis, which determines *XPD* as a function of rain attenuation only. The subject of ice depolarization is addressed in Section 6.2.

6.1.2. Depolarization Measurements

The direct measurement of rain depolarization effects on earth–space paths began in 1975 with the ATS-6 satellite beacons and has continued with CTS, COMSTAR, SIRIO, and other satellites. Several examples of cross-polarization discrimination, *XPD*, as a function of copolarized attenuation, measured on satellite links, are shown in Figures 6-6 through 6-10.

Figure 6-6 shows the *XPD* at 11.6 GHz, circular polarization, elevation angle 33 degrees, measured at Blacksburg, VA for calender year 1978 with the CTS Satellite [6.7]. Three curves are superimposed with the data points. The curve labeled Instantaneous Fit is a least squares fit of the *XPD* and attenuation points for all attenuation values greater than 5 dB. The Statistical Fit curve is based on a comparison of equal probability *XPD* and attenuation from the annual cumulative distributions. The CCIR Fit is based on the prediction given by Equation (6-11) with the CCIR coefficients of Equations (6-12) and (6-13). The CCIR prediction agrees fairly well with the statistical trend of the data, but tends to predict a more severe (lower) *XPD* for a given attenuation, by about 1 dB.

Similar data are shown in Figure 6-7(a,b) for *XPD* measurements on 19 and 28 GHz links at Blacksburg using the COMSTAR satellite. Here the polarization in linear, with a 60° tilt angle. The CCIR Fits are seen to be much closer to the measured data instantaneous fits.

Figure 6-8 shows *XPD* measured at 11.7 GHz at Ottawa, Canada with the CTS over a two month summer period [6.8]. The data are presented on a semi-log plot and the CCIR Fit plots as a straight line. The CCIR Fit agrees with the data set fairly well.

Figure 6-9 presents results of *XPD* measurements over a 12 month period at Austin, Texas, with the CTS satellite [6.9]. The plot shows the mean value of *XPD*, where 50% of the measured points fall above and 50% fall below that value. The 10% and 90% limit curves indicate the range where 10–90% of the *XPD* data points fall for each attenuation value. The CCIR Fits tends to overestimate the severity of the *XPD* by about 3 dB for attenuation values up to 15 dB.

Figure 6-10(a,b) shows *XPD* measured at Holmdel, New Jersey from CTS

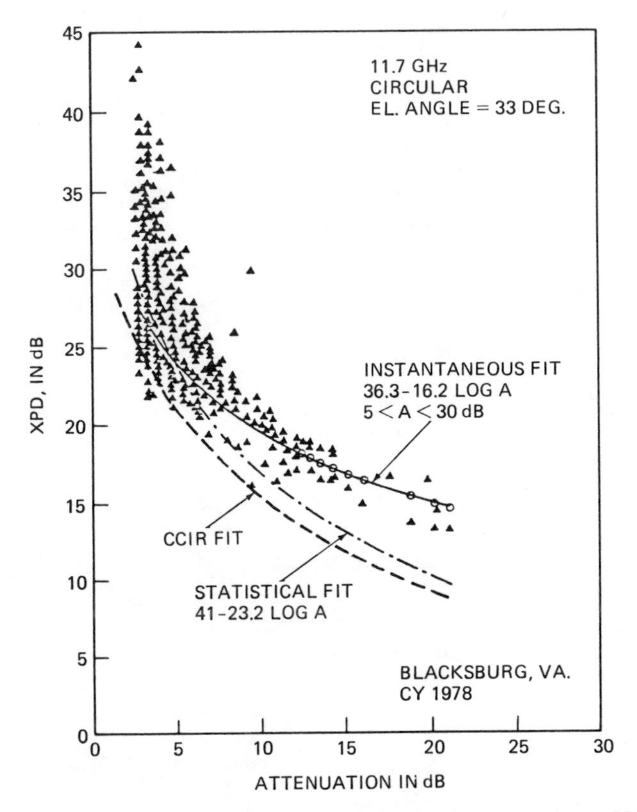

Figure 6-6. Depolarization measurements at 11.7 GHz, Blacksburg, VA.

for several light rain and heavey rain events, respectively [6.10]. The CCIR Fits predict the data quite well even though the sample is small.

The cumulative distribution of *XPD* on an annual or longer term basis is very useful for the evaluation of the effects of depolarization on earth–space paths because it provides a determination of the extent of the problem to be encountered on operational links. Figure 6-11 shows examples of long term cumulative distributions of *XPD* measured at Blacksburg for 11.7 GHz, 19 GHz, and 28 GHz links [6.7, 6.11]. The *XPD* values for 0.01% of the time, corresponding to about 1 hour per year, are about 10–12 dB for 19 and 28 GHz, and about 20 dB for 11.7 GHz. The 11.7 GHz link was circularly polarized and was at a lower elevation angle than the two other links.

The overall *XPD* distributions at 19 and 28 GHz are seen to be very similar. This is not surprising, since a comparison of the *XPD*'s expected from the CCIR coefficients verifies that the difference between the two frequencies is very small.

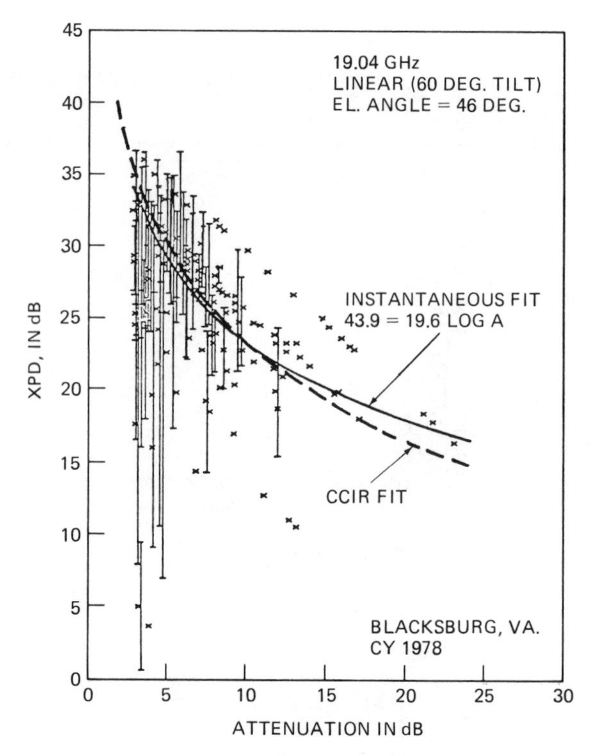

Figure 6-7(a). Depolarization measurements at 19.04 GHz, Blacksburg, VA.

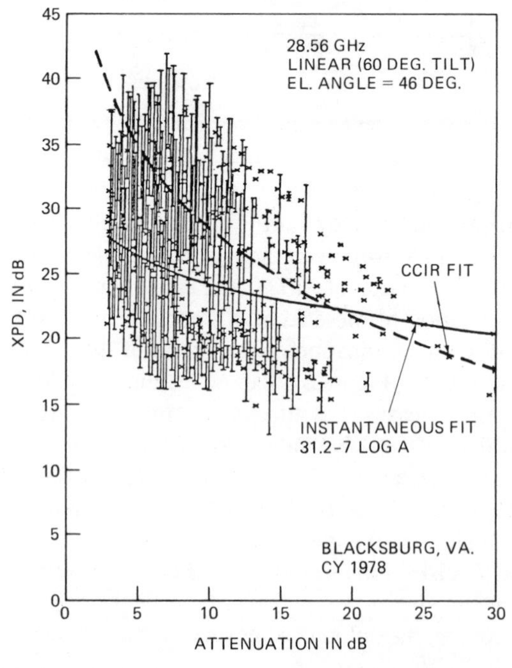

Figure 6-7(b). Depolarization measurements at 28.56 GHz, Blacksburg, VA.

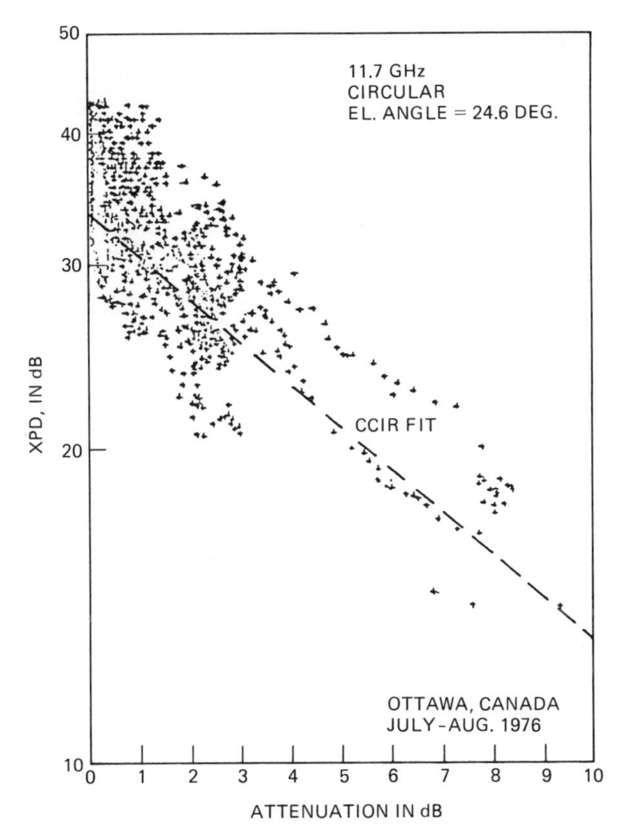

Figure 6-8. Depolarization measurements at 11.7 GHz, Ottawa, Canada.

Consider the difference between the *XPD*'s at each frequency, as determined by Equation (6-11) with the coefficients of Equations (6-12) and (6-13):

$$XPD_{28} - XPD_{19} = 30 \log 28 - 30 \log 19$$
$$-20 \log A_{28} - 20 \log A_{19}$$

where A_{28} and A_{19} are the copolarized attenuations at 28 GHz and 19 GHz, respectively. Using the *a* and *b* coefficients of Table 4-3,

$$XPD_{28} - XPD_{19} = -2.71 + 1.4 \log R \qquad (6\text{-}14)$$

where *R* is the rain rate. For rain rates of 15 mm/h and higher, the difference between *XPD*'s is less than 1 dB, and for lower rain rates the difference is less that about 2 dB. This trend is seen in Figure 6-11, where the 19 and 28 GHz curves are very close at lower *XPD* values (higher rain rates), and diverge slightly at the higher values of *XPD*.

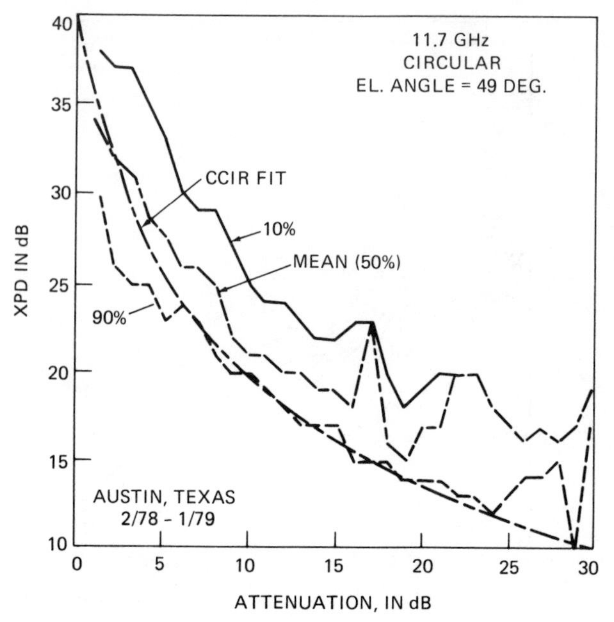

Figure 6-9. Depolarization measurements at 11.7 GHz, Austin, TX.

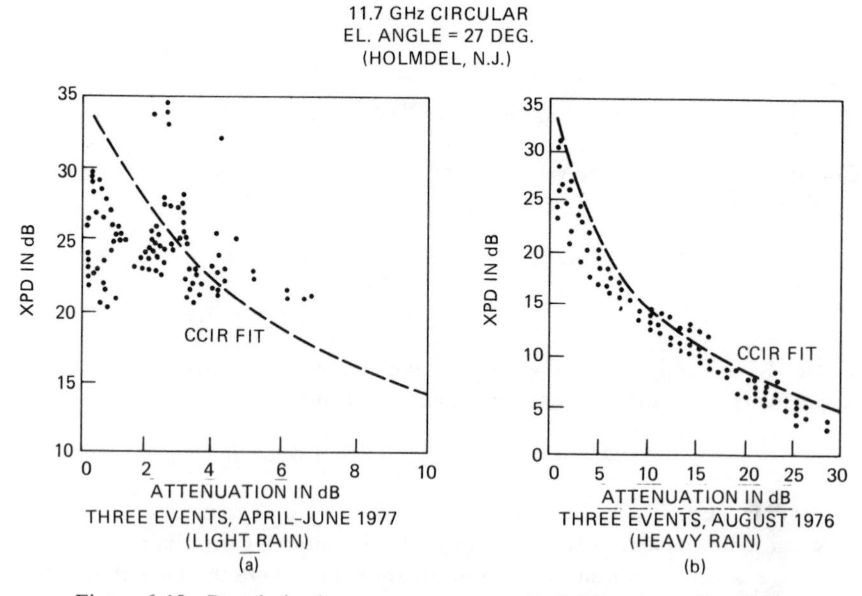

Figure 6-10. Depolarization measurements at 11.7 GHz, Holmdel, NJ.

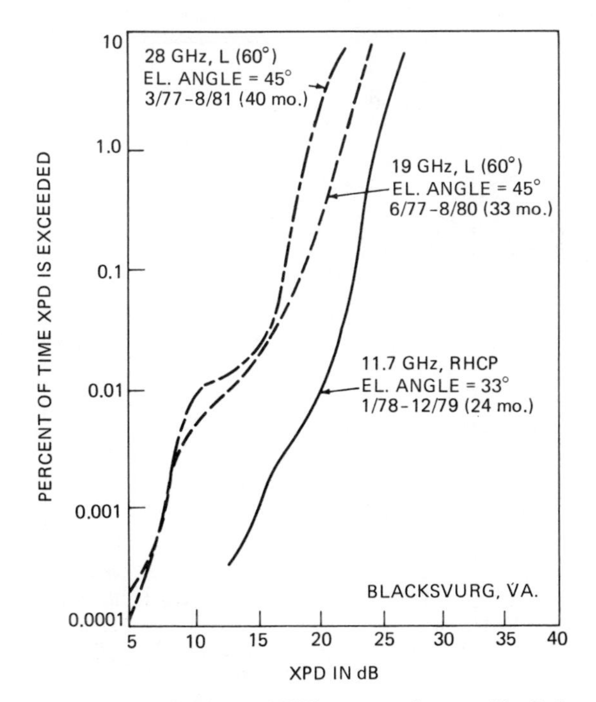

Figure 6-11. Cumulative distributions of *XPD* measured on satellite links at Blacksburg, VA.

Figure 6-12 shows an annual distribution of *XPD* measured on a 19 GHz link at Holmdel, NJ [6.12]. The measurements were for a linear polarized link, with a 69° polarization tilt angle. The *XPD* for 0.01% of the year was 19 dB, which is about 6 dB less severe than similar measurements at Blacksburg, shown in Figure 6-11.

Tables 6-1, 6-2, and 6-3 present summaries of global depolarization statistics in the 4–6 GHz, 12–15 GHz, and 20–30 GHz bands, respectively. The data are presented in descending order of elevation angle in each frequency band. The resultant *XPD* at 0.01% is found to be 20 dB or better for elevation angles above 20° for the 4–6 GHz or 12–15 GHz bands, but drops to 12–18 dB for the 20–30 GHz band. For most systems, a 20 dB *XPD* would be the minimum level for acceptable link performance, with 25–30 dB being a better *XPD* level for high quality links.

6.2. ICE DEPOLARIZATION

Depolarization due to ice crystals has been observed on earth–space paths from measurements with satellite beacons. This condition, initially referred to as "anomalous" depolarization because its cause was unknown, is characterized

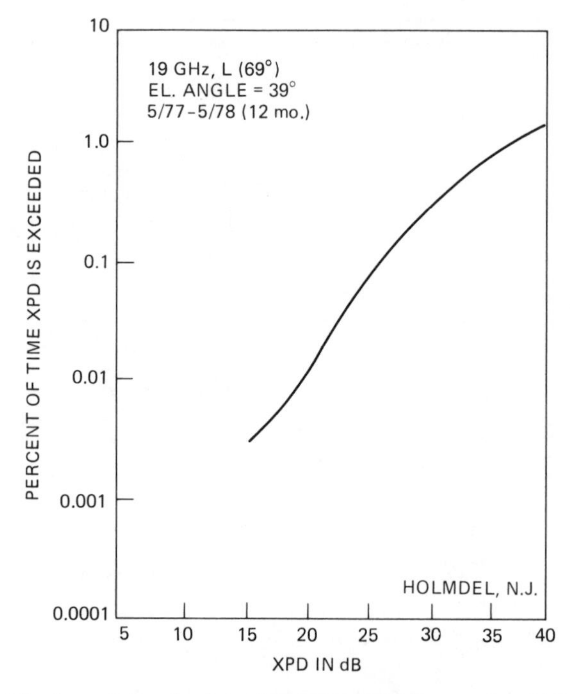

Figure 6-12. Annual distribution of 19 GHz *XPD* measured at Holmdel, Nj.

by a strong depolarization accompanied by very low copolarized attenuation. Also, abrupt changes in the *XPD* have been observed to coincide with lightning discharge in the area of the slant path, suggesting a change in the alignment of the ice crystals.

Ice depolarization occurs when the ice particles are not randomly oriented but have a preferred alignment or direction. Ice crystals most often appear in the shape of needles or plates, and their major axes can abruptly be aligned by an event such as a lightning discharge. When this occurs, differential phase changes in the radiowave can cause an increase in the level of the cross polarized signal level, with little or no change to the copolarized signal. Ice crystals produce nearly pure differential phase shift, without any accompaning differential attenuation, which accounts for the depolarization effects observed in the absence of copolarized attenuation.

6.2.1 Ice Depolarization Measurements

Ice depolarization was observed in early depolarization measurements in the United States using the linearly polarized 20 GHz ATS-6 beacon, but it was not identified as a significant effect until the ATS-6 measurements in Europe a

Table 6-1. Summary of Annual *XPD*. Measurements, 4–6 GHz.

Location	Frequency (GHz)	Elevation Angle	Polarization C, L ($\tau°$)	Time Period	*XPD* at Given Percent of Year		
					0.001%	0.01%	0.1%
Yamaguchi, Japan	4.1	9°	C	8/75–7/77	5	11	16
Sitka, Alaska [6.15]	4.0	11.7°	C	4/80–4/81	18	20.1	25
Taipei, Taiwan [6.14]	4.0	20°	C	3/77–6/78	16	18.3	22
Lario, Italy [6.14]	4.0	24.7°	C	3/77–6/78	20	23.8	30
Baraki, Japan	4.1	35°	C	4/76–3/77	27	32	36
	6.3				25	28	33

SOURCE: Ref. 6.6, except as noted.

Table 6-2. Summary of Annual *XPD* Measurements, 12–15 GHz.

Location	Frequency (GHz)	Elevation Angle	Polarization C, L ($\tau°$)	Time Period	*XPD* at Given Percent of Year		
					0.001%	0.01%	0.1%
Albertslund, Denmark	11.8	26.5°	C	1/79–12/79	17	23	>28
	14.5				13	18	26
Martlesham, UK	11.6	29.9°	L(8°)	7/78–7/80	28	30	—
	11.8		C		19	22	—
	14.5				17	20	—
	11.8				18.7	24	—
Gometz-la-Ville, France	11.6	32°	C	11/77–11/78	17	22	27
Leeheim, FRG	11.6	32.9°	C	1/79–12/80	14	20	28
Blacksburg, VA	11.7	33°	C	1/78–12/78	14	20	22
				1/79–7/79	16	21	24.5
Gometz-la-Ville, France	11.8	33.6°	C	1/79–11/79	16	22	28
Kashima, Japan	11.5	47°	C	5/77–4/78	—	24	29
Austin, TX	11.7	49°	C	6/76–6/79	15	20.5	28

SOURCE: Ref. 6.6.

Table 6-3. Summary of Annual *XPD* Measurements, 20–30 GHz.

Location	Frequency (GHz)	Elevation Angle	Polarization C, L ($\tau°$)	Time Period	*XPD* at Given Percent of Year		
					0.001%	0.01%	0.1%
Holmdel, NJ	19	39°	L(21°)	5/77–5/78	>10	16	24
	19		L(69°)	5/77–5/78	>10	18	25
	28.6		L(69°)	5/77–5/78	—	12	22
Blacksburg, VA	28.6	45°	L (60°)	1/78–12/78	9.5	13	15
				1/79–6/79	7	12.5	18
Kashima, Japan	34.5	47°	C	5/77–4/78	—	17	22
	19.5	48°	C	8/79–7/80	—	20	26
Yokohama, Japan	19.5	48°	C	4/79–3/80	—	17	21

SOURCE: Ref. 6.6.

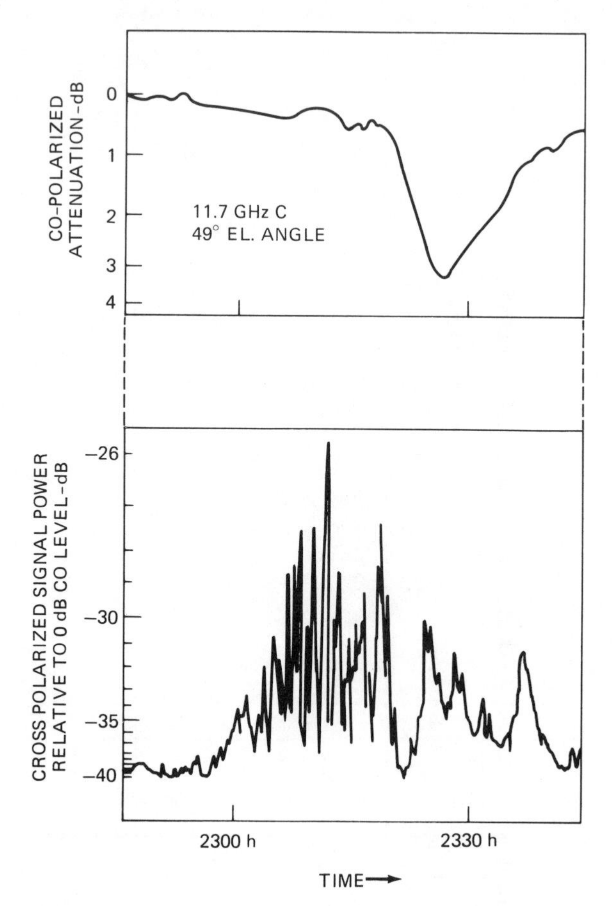

Figure 6-13. Ice depolarization event at Austin, TX, 11.7 GHz, 49° elevation angle, June 26, 1976.

year later, and the CTS and COMSTAR measurements in the United States which followed [6.13]. Ice depolarization effects have been observed in the U.S., Europe and Japan, at frequencies from 4 GHz to 30 GHz [6.13–6.18].

Figure 6-13 shows an example of an ice depolarization event observed at Austin, Texas with the 11.7 GHz CTS beacon [6.9]. The cross-polarized signal level begins to increase at 2300 h, reaching a peak change of about 14 dB, while the co-polarized attenuation remains less than 1 dB.

Figure 6-14 shows another example of ice depolarization, observed at Slough, England with the 30 GHz ATS-6 beacon [6.19]. The first event, beginning at about 2338 h, occured during intense lightning, and the next two events occured during rain, as indicated by the notes of the plot. The *XPD* degraded by over 20 dB during the intense lightning, with copolarized attenuation remaining less

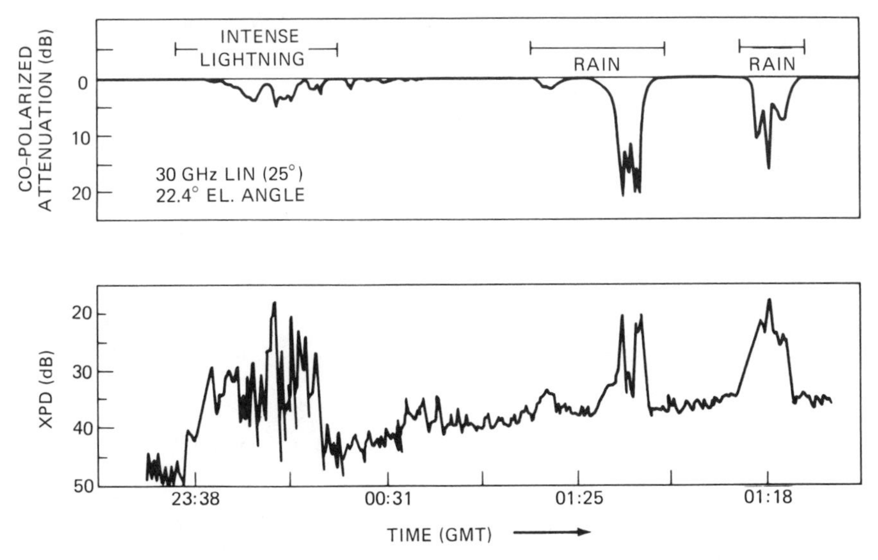

Figure 6-14. Ice depolarization observed at Slough, England, 30 GHz, 22.4° elevation angle, July 15–16, 1976.

than 5 dB, and with little time correlation between the two. The *XPD* variations during the rain periods, on the other hand, are seen to be well time correlated with the copolarized attenuation variations.

These examples are typical of ice depolarization events observed on satellite paths. Ice depolarization often (but not always!) occurs several minutes before the appearence of a severe rain attenuation event. Ice depolarization characteristics have been observed in the presence of clouds, light precipitation, and in clear sky, as well as during the occurence of lightning discharges.

The contribution of ice depolarization to the total depolarization on a radiowave link is difficult to determine from direct measurement, but can be inferred from observation of the copolarized attenuation during depolarization events. The depolarization which occurs when the copolarized attenuation is low, (i.e., less than 1–1.5 dB), can be assumed to be caused by ice particles alone, while the depolarization which occurs when copolarized attenuation is higher can be attributed to both rain and ice particles.

The relative occurrence of ice depolarization versus rain depolarization is observed in Figure 6-15, which shows the three year cumulative distributions of *XPD* measured at Austin, Texas at 11.7 GHz [6.9]. The curve for rain and ice is for all of the observed depolarization, while the rain only curve corresponds to those measurements where the copolarized attenuation was greater than 1 dB. The ice contribution is observable only for *XPD* values greater than 25 dB, where ice effects are present about 10% of the time. The ice effects

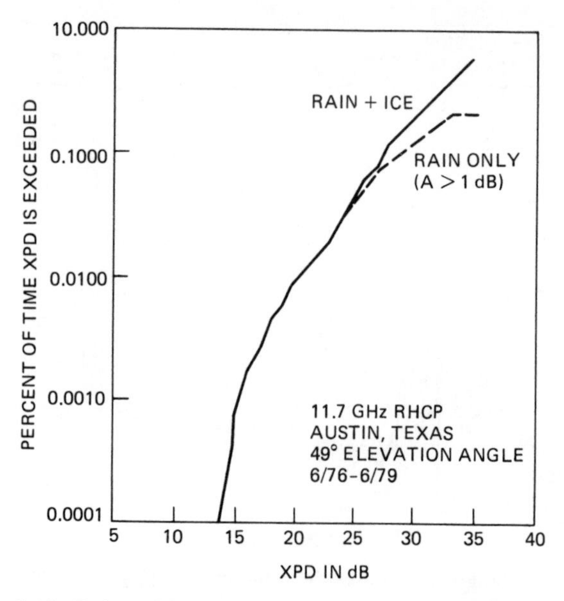

Figure 6-15. Rain and ice depolarization measurements at 11.7 GHz.

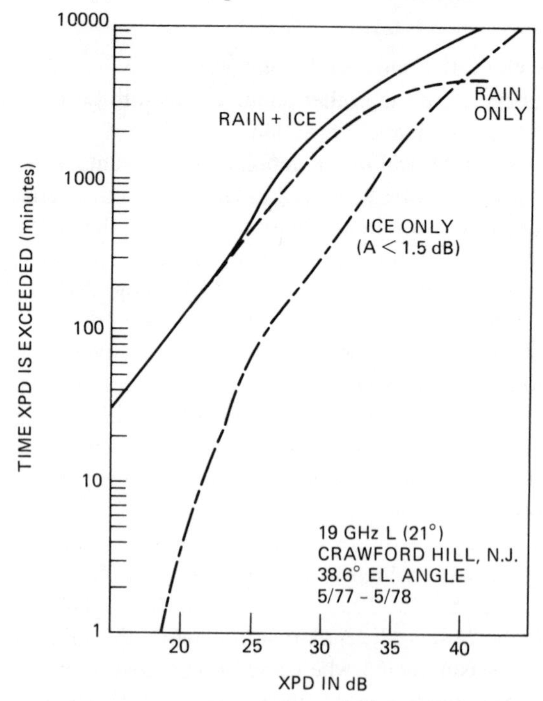

Figure 6-16. Rain and ice depolarization measurements at 19 GHz.

exceed 50% of the time at an *XPD* of 35 dB. The decrease in *XPD* caused by ice averages about 2–3 dB for a given percent of time.

Similar results were observed at 19 GHz from measurements reported at Crawford Hill, New Jersey, using the COMSTAR beacons, as shown on Figure 6-16 [6.12]. For these measurements, ice depolarization was assumed for all times when the copolarized attenuation was 1.5 dB or less. Curves for rain and ice, ice only, and rain only, are shown. Again, the ice contribution is observable only for *XPD* values of 25 dB or greater, corresponding to an annual availibility of about 0.063% for this data set. The decrease in *XPD* due to ice for a given time ranges from about 1 to 5 dB over that range.

6.2.2 Ice Depolarization Prediction

The development of an ice depolarization model for radiowave links is hampered by the lack of definitive information on the size, shape, and density of ice crystals in the path. The crystals most often appear in the shape of needles or plates, which are approximated by a long, narrow prolate spheroid or a flat oblate spheroid, respectively. Forward scattering from the ice crystals is calculated by assuming the Rayleigh approximation (see Section 4.1), since the particles are small in comparison to the wavelength.

Several attempts [6.13, 6.20, 6.21] have been initiated in the calculation of ice particle depolarization on earth–satellite paths; however, a definitive prediction procedure, such as the CCIR method for rain depolarization, is not yet available. It does seem apparent in the theoretical developments to date, though, that ice particle depolarization is minor compared to rain induced depolarization, and this result is substantiated by the measurements described in the previous section.

Stutzman et al. [6.22] concluded, on the basis of depolarization measurements in the 11 GHz region, that ice depolarization apears "to have negligible statistical significance, except for, at most, a 2 dB degradation at low elevation angles."

The CCIR suggests in Report 564-2 that a constant factor be subtracted from the *XPD* calculated by the CCIR rain depolarization prediction procedure (described in Section 6.1.1) to allow for the contribution of ice crystals [6.6]. A value of 2 dB "seems to be reasonable for North America," according to that report, while a value of 4 or 5 dB "may be more appropriate" for the maritime climate of Northwestern Europe. A later CCIR document [6.23] presents the same values and in addition suggests that the ice depolarization contribution can be ignored for time percentages less than 0.1%.

6.2.3 Summary

Measurements have shown that ice depolarization is *statistically* not a major contributor to the depolarization experienced on earth–space paths. Ice depo-

larization is only measurable for fairly large values of *XPD*, i.e., 25 dB or greater, where link performance is generally not adversly affected. The reduction in *XPD* caused by ice effects is small, on the order of 2–4 dB for a given percent of time. For lower values of *XPD*, where link performance is critically dependent on *XPD*, rain depolarization predominates.
Individual events of ice particle induced *XPD* variations can be severe, however, and may have to be accounted for in systems which are sensitive to abrupt changes in the cross-polarized phase and amplitude components of the transmitted signal.

6.3. MULTIPATH DEPOLARIZATION

Depolarization effects have been observed on terrestrial communications links during deep copolarized signal fading associated with multipath conditions on the link. Multipath conditions result in the transmitted radiowave reaching the receiving antenna by two or more propagation paths. During these clear weather conditions the *XPD* can degrade severely from the selective fading of the copolarized (desired) signal and the crosspolarized (undesired) signal, resulting in cancellation and enhancement at the receiving antenna.

Figure 6-17 shows an example of 11.6 GHz signal level variations on a 42.5 km terrestrial path during both non-fading and fading periods [6.24]. On the upper plot, (a), which shows a typical nonfading hour, the co- and cross-polarized signal levels show some scintillations but the *XPD* remains at the nomial 20 dB level. On the lower plot, (b), during an hour of severe multipath fading, the *XPD* is seen to be degraded, reaching a value of 0 dB at one time during the hour. There was no rain on the path during these observations.

Several possible depolarization mechanisms can produce the effects described above. They involve contributions from both the propagation medium itself and the cross-polarization patterns of the receive and transmitt antennas [6.25].

Those mechanisms involving only the propagation medium include:

1. Depolarization of the direct copolarized signal by tropospheric turbulence along the path.
2. Depolarization of an indirect component of the copolarized signal due to reflection from an atmospheric layer.
3. Depolarization of an indirect component of the copolarized signal due to scattering or reflection from land or water surfaces along the path.
4. Depolarization of a direct component of the wave due to refractive bending.

Mechanisms which involve additional effects caused by antenna patterns include:

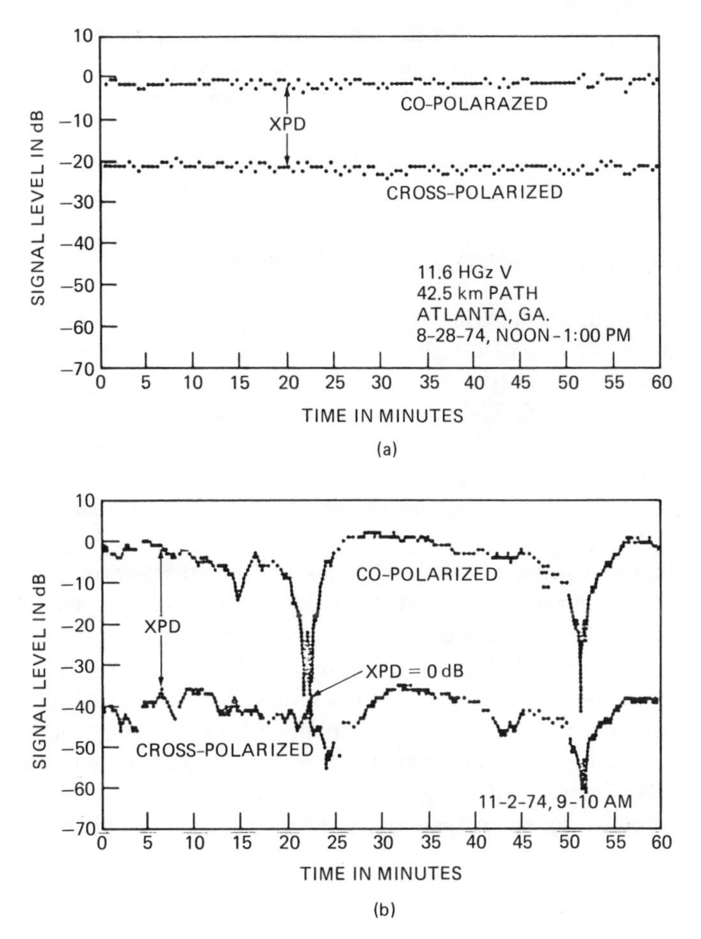

Figure 6-17. Multipath depolarization on an 11.6 GHz terrestrial link.

5. Coupling of an indirect component of the signal reflected from (or refracted through) an atmospheric layer via the cross-polarized pattern(s) of the antennas.
6. Coupling of an indirect component of the signal via the cross-polarized antenna pattern(s) due to reflection from land or water surfaces.
7. Coupling of an indirect component of the signal via the cross-polarized antenna pattern(s) due to multiple reflections between the ground and the atmosphere.

All of these mechanisms could occur to some extent during multipath fading; however, it is expected that one or two would predominate during extremely severe reductions in the *XPD*.

It should be emphasized that for most typical earth–space communications links, which employ very directive (i.e., narrow beamwidth) ground antennas, multipath conditions will generally not be significant except for very low elevation angles (5–10° or less). The exception would be for low gain antenna mobile communications links operating in the bands below 3 GHz, where multipath conditions will occur often and can be very severe at any elevation angle.

6.3.1. Multipath Depolarization Characteristics

Virtually no information exists on the measurement and evaluation of multipath depolarization effects on slant paths. The measurements and models that have been developed for terrestrial links, while not directly applicable, do offer some insight into the problem, and they will be discussed here.

For deep copolarized fade depths (i.e., 15 dB or higher), the rms value of *XPD* tends to decrease linearly with copolarized fade depth. Figure 6-18 shows an example of this type of behavior for the 11.6 GHz link in Georgia discussed previously [6.26].

This behavior can be modeled by describing the crosspolarization interference as the sum of two components. The first component is dependent on copolarized signal level. The second component represents the random received signal which is Rayleigh distributed and independent of the copolarized signal level. The resultant *XPD* level is then described by a Rice–Nakagami distribution, which reduces to a Rayleigh distribution for large fade depths.

Several prediction methods for multipath induced depolarization on terrestrial links have been developed using similar modeling concepts [6.24–6.28]. The results can be reduced to an approximate relationship for the *XPD* of the form

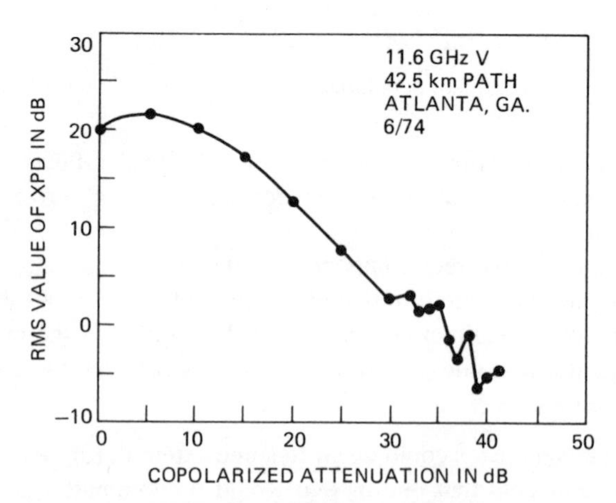

Figure 6-18. Dependence of *XPD* on copolarized attenuation during multipath fading.

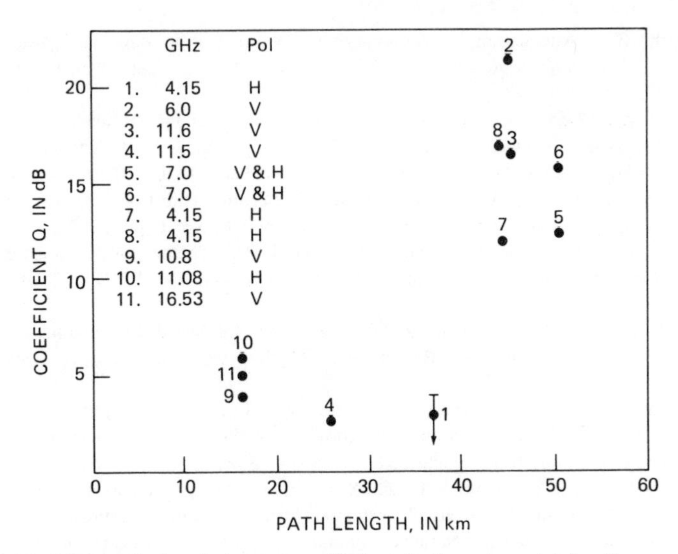

Figure 6-19. Multipath depolarization coefficient Q, from terrestrial path measurement.

$$XPD \approx XPD_0 - A + Q \tag{6-15}$$

where A is the copolarized attenuation, in dB, XPD_0 is the static value of XPD during nonfading conditions, and Q is an empirically determined coefficient developed from long term equiprobable values of XPD and A. There is some evidence that Q is inversely related to the slope to the vertical cross-polarization antenna pattern [6.25].

Figure 6-19 summarizes values of Q as a function of path length, derived from global XPD measurements reported by the CCIR [6.25]. The values vary over a large range, probably due to the differences in the cross-polarization characteristics of the link antennas more that any other factor. The coefficient tends to be higher for longer path lengths, indicating less of a degradation on XPD over the longer distances, where multipath variations may be averaged out.

The application of these results to earth–space slant paths, even at very low elevation angles, is difficult, since the path length appears to be a critical factor, along with the specific polarization patterns of the ground antennas. Further information is required, particularly directly measured data, to fully access the impact of multipath depolarization on slant paths.

REFERENCES

6.1. Pruppacher, H. R., and Pitter, R. L., "A Semi-Empirical Determination of the Shape of Cloud and Rain Drops," *Journal of the Atmospheric Sciences*, Vol. 28, pp. 86–94, January 1971.

6.2. Oguchi, T., ''Attenuation of Electromagnetic Waves Due to Rain with Distorted Raindrops,'' *Journal of the Radio Research Lab.*, Vol. 7, No. 33, Sept. 1960; Part II, Vol. 11, No. 53, Jan. 1964.

6.3. Chu, T. S., ''Rain-Induced Cross-Polarization at Centimeter and Millimeter Wavelengths,'' *The Bell System Technical Journal*, Vol. 53, No. 8, pp. 1557–1579, October 1974.

6.4. Arnold, H. W., Cox, D. C., Hoffman, H. H., and Leck, R. P., ''Characteristics of Rain and Ice Depolarization for a 19- and 28-GHz Propagation Path from a Comstar Satellite,'' *IEEE Trans. on Antennas and Propagation*, Vol. AP-28, No. 1, pp. 22–28, Jan. 1980.

6.5. Nowland, W. L., Olsen, R. L., and Shkarofsky, I. P., ''Theoretical Relationship Between Rain Depolarization and Attenuation,'' *Electronics Letters*, Vol. 13, No. 22, pp. 676–678, October 27, 1977.

6.6. CCIR, Report 564-2, ''Propagation Data Required for Space Telecommunications Systems,'' Recommendations and Reports of the CCIR—1982, Volume V, pp. 185–194, Geneva, 1982.

6.7. Bostian, C. W., ''A Depolarization and Attenuation Experiment using the COMSTAR and CTS Satellites,'' Final Report on the Fourth Year of Work, NASA Contract NAS5-22577, Virginia Polytechnic Institute and State University, Blacksburg, VA, March 25, 1980.

6.8. Nowland, W. L., Stickland, J. I., and Schlesak, J., ''Measurements of Precipitation and Depolarization Using the 12 GHz Transmissions from the Communications Technology Satellite Hermes,'' 1976 U.S. National Committee Meeting of URSI, U. of Massachusetts, Amherst, Mass., October 11–15, 1976.

6.9. Vogel, W. J., ''CTS Attenuation and Cross Polarization Measurements at 11.7 GHz,'' Final Report, Report No. 22576-1, June 1980.

6.10. Rustako, Jr., ''An Earth–Space Propagation Measurement at Crawford Hill Using the 12 GHz CTS Satellite Beacon,'' *The Bell System Technical Journal*, Vol. 57, No. 5, May–June 1978.

6.11. Andrews, J. H. et al., ''Results of the VPI&SU COMSTAR Experiment,'' *Radio Science*, Vol. 17, No. 6, pp. 1349–1359, Nov.-Dec. 1982.

6.12. Cox, D. C., and Arnold, H. W., ''Results from the 19- and 28-GHz COMSTAR Satellite Propagation Experiments at Crawford Hill,'' *Proc. of the IEEE*, Vol. 70, No. 5, May 1982.

6.13. Bostian, C. W., and Allnutt, ''Ice Crystal Depolarization on Satellite–Earth Microwave Radio Paths,'' *IEE*, Vol. 126, No. 10, pp. 951–960, Oct. 1979.

6.14. Kennedy, D. J., ''Rain Depolarization Measurements at 4 GHz,'' *COMSAT Tech. Review*, Vol. 9, No. 2B, pp. 629–668, Fall 1979.

6.15. Struharik, S. J., ''Rain and Ice Depolarization Measurements at 4 GHz in Sitka, Alaska,'' *COMSAT Tech. Review*, Vol. 13, No. 2, pp. 403–435, Fall 1983.

6.16. Stutzman, W. L., Bostian, C. W., Tsolakis, A., and Pratt, T., ''The Impact of Ice Along Satellite-to-earth Paths on 11 GHz Depolarization Statistics,'' *Radio Science*, Vol. 18, No. 5, pp. 720–724, Sept.-Oct. 1983.

6.17. Goldhirsh, J., ''Ice Depolarization of the COMSTAR Beacon at 28.56 GHz During Low Fades and Correlation with Radar Backscatter,'' *IEEE Trans. on AP*, Vol. AP-30, No. 2, pp. 183–190, Mar. 1982.

6.18. Fujita, M., et al, ''ETS-II Experiments Part IV: Characteristics of Millimeter and Centimeter Wavelength Propagation,'' *IEEE Trans. on AES*, Vol. AES-16, No. 5, pp. 581–589, Sept. 1980.

6.19. Shutie, P. F., Allnut, J. E., and Mackenzie, E. C., ''Depolarization Results at 30 GHz Using Transmissions from the ATS-6 Satellite,'' URSI Commission F, Open Symposium on Propagation in Non-Ionized Media, La Baule, France, Apr. 28–May 6, 1977.

6.20. Evans, B. G., and Holt, A. R., ''Cross Polarization Phase Due to Ice Crystals on Microwave Satellite Paths,'' *Electronics Letters*, Vol. 13, No. 22, pp. 664–666, Oct. 27, 1977.

6.21. Tsolaskis, A., and Stutzman, W. L., ''Calculation of Ice Depolarization on Satellite Radio

Paths,'' *Radio Science*, Vol. 18, No. 6, pp. 1287–1293, Nov.-Dec. 1983.
6.22. see [6.16].
6.23. CCIR Document 5/10, ''Interim Working Party 5/2—Propagation Data Required for Space Telecommunications Systems,'' CCIR Study Group 5 Documents, Geneva, Mar. 30, 1983.
6.24. Lin, S. H., ''Impact of Microwave Depolarization During Multipath Fading on Digital Radio Performance,'' *Bell System Tech. Journal*, Vol. 56, No. 5, pp. 645–674, May 1977.
6.25. CCIR Report 722-1, ''Cross-Polarization Due to the Atmosphere,'' Volume V, Recommendations and Reports of the CCIR—1982, pp. 185–194, Geneva, 1982.
6.26. Mottl, T. O., ''Dual-Polarized Channel Outages During Multipath Fading,'' *Bell System Technical Journal*, Vol. 56, No. 5, May–June 1977.
6.27. Morita, K., et al, ''A Method for Estimating Cross-Polarization Discrimination Ratio During Multipath Fading,'' (in Japanese), *Trans. Inst. Electron Comm. Engrs.*, Japan, Vol. 62-B, pp. 998–1005, 1979.
6.28. Olsen, R. L., ''Cross Polarization During Clear-air Conditions on Terrestrial Links: A Review,'' *Radio Science*, Vol. 16, No. 5, pp. 631–647, Sept.-Oct. 1981.

CHAPTER 7
RADIO NOISE IN SATELLITE COMMUNICATIONS

There are several natural and man-made sources of unwanted external noise which can be introduced in the radiowave transmission of a space communications link. Any natural absorbing medium in the atmosphere which interacts with a radiowave will not only produce a signal amplitude reduction (attenuation), but will also be a source of thermal noise power radiation. The noise associated with these sources, referred to as radio noise, or sky noise, will directly add to the system noise through an increase in the antenna temperature of the receiver. For very low noise communications receivers, such as those in the NASA deep space tracking network, radio noise can be the limiting factor in the design and performance of the system.

Radio noise is emitted by all matter, both terrestrial and extraterrestrial. Terrestrial sources are both *natural*, such as gaseous atmospheric constituents (oxygen and water vapor), and hydrometeors (clouds and rain), and *man-made*, such as emission from electric devices and from other communications systems.

Figure 7-1 pictorally displays the major natural sources of radio noise which could be present on the downlink of a space communications link. The noise radiation can enter the ground receiver through either the main beam or the sidelobes of the antenna, therefore the source need not be in direct line of sight to the satellite.

Figure 7-2 shows a similar display for the uplink. Here, the uplink emission of the atmospheric constituents will usually fill a small part of the satellite antenna beam, while surface emissions will fill the major part. The total uplink noise temperature, as seen from space, can range from 140 to 290 degrees Kelvin (°K), depending on the percentage of land versus sea in the antenna beam, the degree of cloud cover, and other link factors.

Figure 7-3 summarizes the minimum expected noise levels produced by sources of external radio noise in the frequency range applicable to practical space communications [7.1]. The noise is expressed in terms of an equivalent noise temperature t_a, in °K, and a noise factor, F_a, in dB, given by

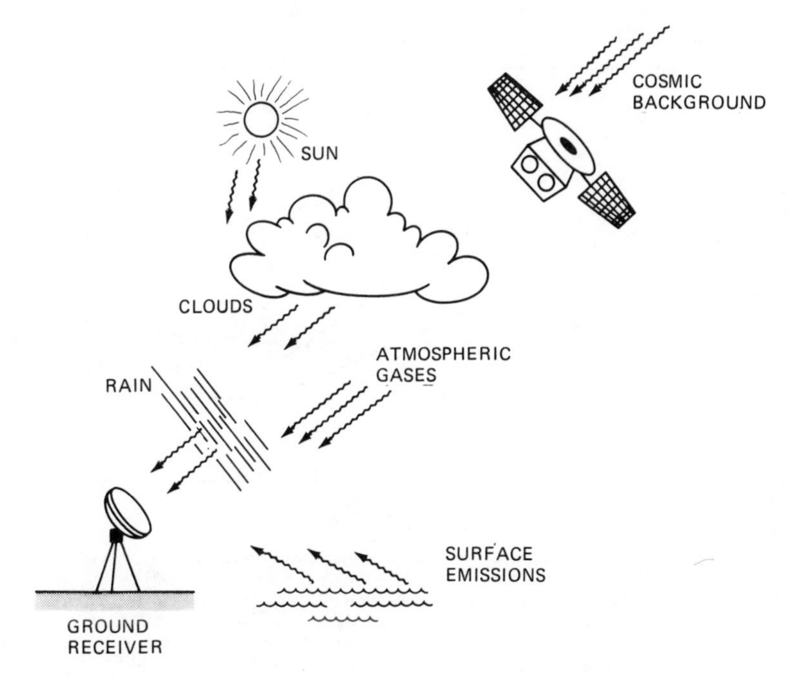

Figure 7-1. Natural sources of radio noise on the downlink of a space communications link.

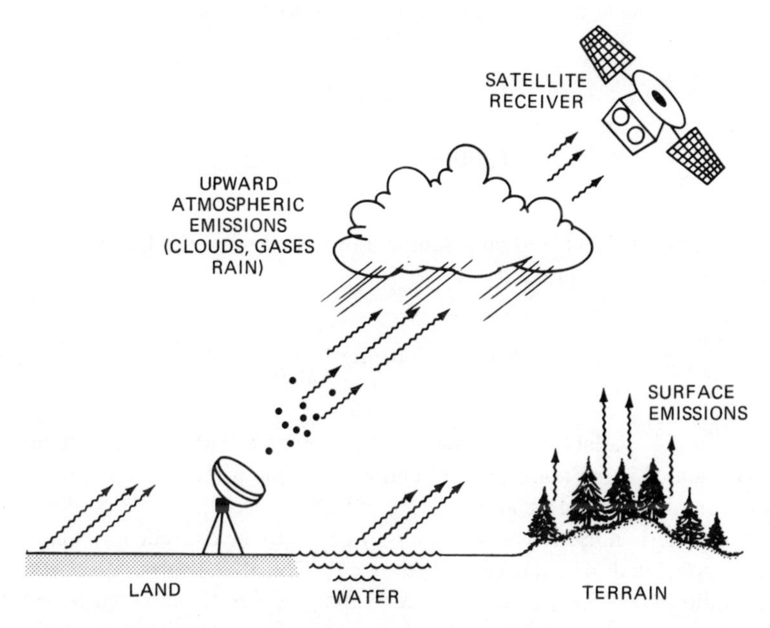

Figure 7-2. Natural sources of a radio noise on the uplink of a space communications link.

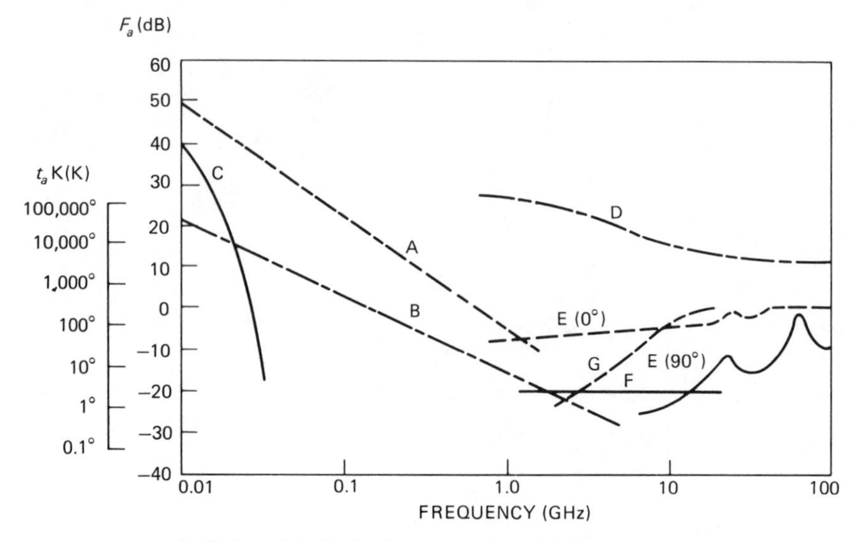

A: Estimated median business area man-made noise
B: Galactic noise
C: Atmospheric noise, value exceeded 0.5% of time
D: Quiet sun (½ degree beamwidth directed at sun)
E: Sky noise due to oxygen and water vapor (very narrow beam antenna); upper curve, 0° elevation angle; lower curve, 90° elevation angle
F: Black body (cosmid background), 2.7 K
G: Heavy Rain (50 mm/h over 5 km)

Figure 7.3. Minimum expected external noise from natural and man-made sources, 10 MHz to 100 GHz.

$$F_a(\text{dB}) = 10 \log \left(\frac{t_a}{t_0}\right) \tag{7-1}$$

where t_0 is the ambient reference temperature, set to 290°K. Equivalently, the noise factor can be expressed as

$$F_a(\text{dB}) = 10 \log \left(\frac{p_m}{kt_0B}\right) \tag{7-2}$$

where p_m is the available noise power at the antenna terminals, k is Boltzmann's constant, and B is the noise power bandwidth of the receiving system.

Between about 30 MHz and 1 GHz, galactic noise (curve B) predominates over atmospheric noise (curve C), but will generally be exceeded by man-made noise in populated area (curve A). Above 1 GHz, the absorptive constituents of the atmosphere, i.e., oxygen, water vapor (curves E) and rain (curve G), also act as noise sources and can reach a maximum value of 290°K under extreme conditions.

The Sun is a strong variable noise source, reaching values of 10,000°K and higher when observed with a narrow beamwidth antenna (curve D), under quiet Sun conditions. The cosmic background noise level of 2.7°K (curve F) is very low and is not a factor of concern in space communications.

The remainder of this chapter discusses the noise contributions of the natural sources of external radio noise, i.e., gaseous constituents, clouds, rain, earth surface emissions, and extra-terrestrial sources.

7.1. NOISE FROM ATMOSPHERIC GASES

The gaseous constituents of the Earth's atmosphere interact with a radiowave through a molecular absorption process which results in attenuation of the wave (see Chapter 3, Section 3.1). This same absorption process will produce a thermal noise power radiation which is directly related to the intensity of the absorption.

The effective sky noise temperature T_s for a downlink ground receiver, caused by an absorbing medium, is quantitatively described from application of the equation of radiative transfer by

$$T_s = \int_0^{\infty} T(s)\, \gamma(s)\, e^{-\int_0^s \gamma(s')ds'}\, ds \tag{7-3}$$

where $T(s)$ is the physical temperature of the medium, in °K, $\gamma(s)$ is the absorption coefficient of the medium, in km, and s is the distance along the path, from the antenna, in km.*

If $T(s)$ is replaced by a mean path temperature T_m, the above simplifies to

$$T_s = T_m\left(1 - \frac{1}{L}\right) \tag{7-4}$$

where L is the loss factor due to the absorbing medium. If L is expressed in dB, then

$$T_s = T_m(1 - 10^{-[A(\mathrm{dB})/10]}) \tag{7-5}$$

If a horizontally stratified atmosphere is assumed, T_s at an elevation angle θ is given by (see Appendix A)

*The cosmic background noise contribution is not included in this expression.

$$T_s(\theta) = T_m(1 - 10^{-[A_z(\text{dB})/10\sin\theta]}), \qquad \theta \geq 10° \tag{7-6}$$

where A_z (dB) is the attenuation in the zenith ($\theta = 90°$) direction.

The major atmospheric gases that affect space communications, as described in Chapter 3, are oxygen and water vapor. The sky noise temperature for oxygen and water vapor, for an infinitely narrow beam, at various elevation angles, was calculated by direct application of the radiative transfer equation, at frequencies between 1 and 340 GHz [7.2, 7.3]. Figures 7-4, 7-5, and 7-6 summarize the results of these calculations for three representative atmospheric conditions. Figure 7-4 represents a dry atmosphere (3 g/m^3 water vapor, 17% relative humidity); Figure 7-5 represents a moderate atmosphere (7.5 g/m^3, 50% relative humidity); and Figure 7-6 a very humid atmosphere (17 g/m^3, 98% relative humidity).

Figure 7-7 provides an expanded frequency scale version of the moderate atmosphere case (Figure 7-5) for use with frequencies below 60 GHz.

A horizontally stratified atmosphere, corrected for Earth curvature, is as-

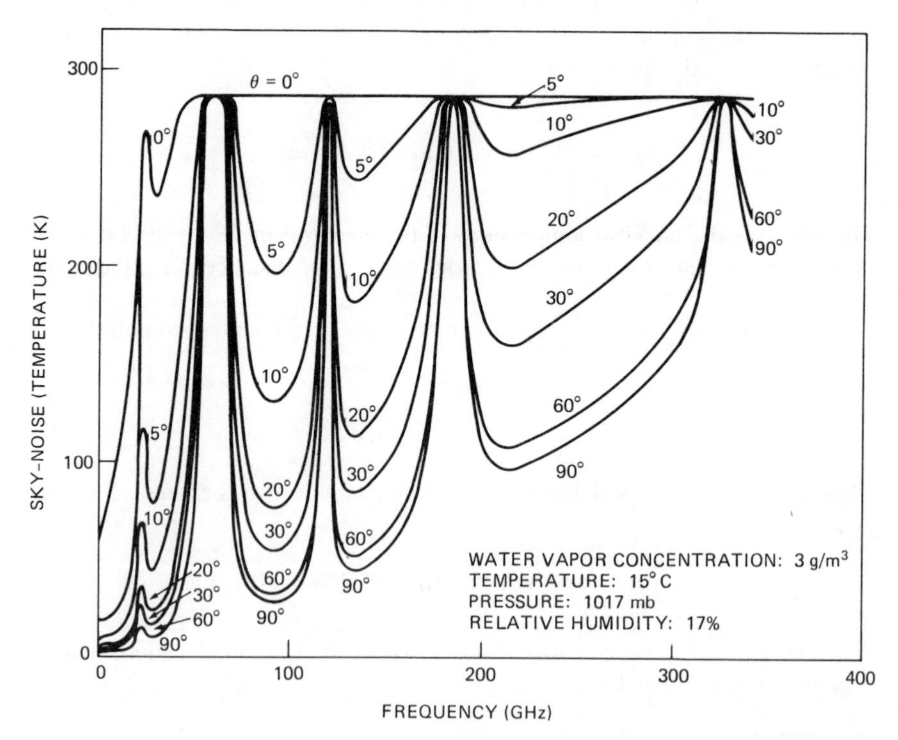

Figure 7-4. Sky noise temperature for a *dry* atmosphere. θ is the elevation angle.

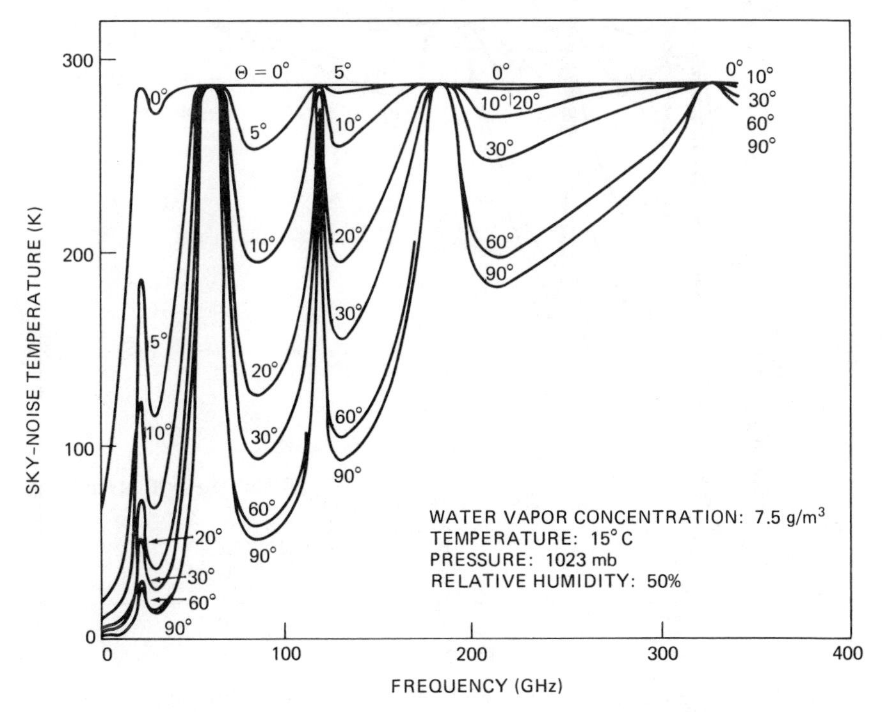

Figure 7-5. Sky noise temperature for a *moderate* atmosphere. θ is the elevation angle.

sumed for the calculations. An exponential decrease of water vapor with height is assumed, with a scale height of 2 km.

The sky temperature at a 30° elevation angle is seen to be less than 10°K for all three cases at 12 GHz, and ranges from 18°K to 50°K at 20 GHz, and 17°K to 42°K at 30 GHz. For a receiver system with a 3 dB (290°K) noise figure front end, the contribution of the atmospheric sky noise will be small, i.e., less than 0.4 dB for all of the examples cited above.

The results given in the above figures at frequencies below 60 GHz are in agreement to within about 15% with the limited amount of experimental data available [7.3].

7.2. NOISE FROM CLOUDS

Sky noise from clouds can be determined from radiative transfer approximations in much the same way as given in the previous section for atmospheric gases. The temperature and cloud absorption coefficient variations along the path must be defined, and Equations (7-3) through (7-6) can be applied.

A detailed study of the effects of clouds on radiowave paths was accom-

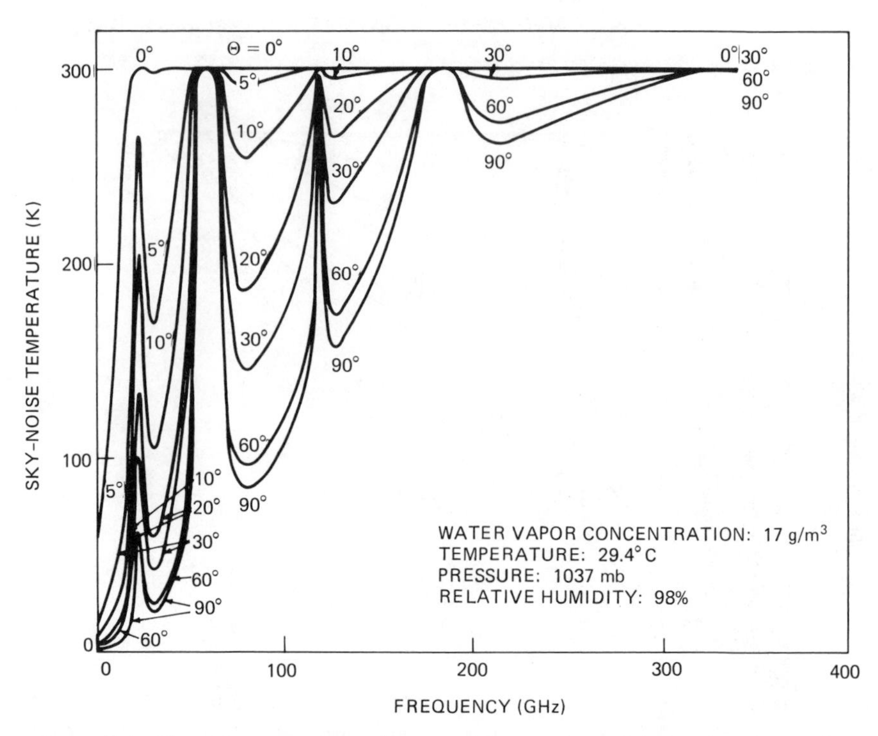

Figure 7-6. Sky noise temperature for a *humid* atmosphere. θ is the elevation angle.

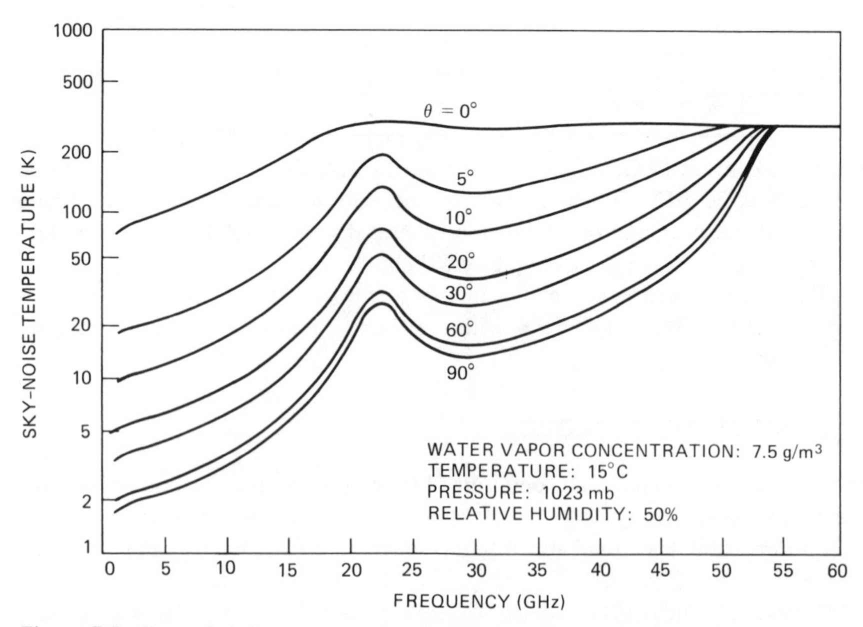

Figure 7-7. Expanded frequency scale plot of sky noise temperature for a moderate atmosphere.

Table 7-1. Zenith Sky Temperature from the Slobin Cloud Model.

Frequency (GHz)	Light, Thin Cloud	Light Cloud	Medium Cloud	Heavy Clouds I	Heavy Clouds II	Very Heavy Clouds I	Very Heavy Clouds II
6/4	<6°	<6°	<13°	<13°	<13°	<19°	<19°
14/12	6	10	13	19	28	36	52
17	13	14	19	28	42	58	77
20	16	19	25	36	52	77	95
30	19	25	30	56	92	130	166
42	42	52	68	107	155	201	235
50	81	99	117	156	204	239	261

plished by Slobin [7.4]. Cloud attenuation and noise temperature calculations were made for several locations in the United States, using radiative transfer methods and a four layer cloud model. Details of the model are described in Section 4.3.1 on cloud attenuation prediction.

Table 7-1 summarizes the zenith (90° elevation angle) sky noise temperature as calculated by the Slobin model for several frequencies of interest. Values for other elevation angles can be determined by the procedures described in Appendix A. For example, for a 20 GHz link at a 30° elevation angle, the sky temperature due to a medium cloud will be

$$\frac{25^\circ\text{K}}{\sin 30^\circ} = 50^\circ\text{K}$$

Figure 7-8 (a–d) shows examples of zenith sky temperature cumulative distributions for four of the Slobin cloud regions (Denver, New York, Miami, and Oakland) at frequencies of 10, 18, 32, 44, and 90 GHz. Plots for all fifteen cloud regions are available in Reference 7.4.

The distributions give the percent of the time the noise temperature is the given value or less. For example, at Denver, the noise temperature was 12°K or less for 0.5 (50%) of the time at 32 GHz. Values of noise temperature in the distribution range 0 to 0.5 (0 to 50%) may be regarded as the range of clear sky conditions. The value of noise temperature at 0% is the lowest value observed for the test year.

Sky temperature values can be approximated from cloud attenuation values by use of Equation (7-5) with T_m, the mean path temperature for clouds, set to 280°K.

7.3. NOISE FROM RAIN

Sky noise due to absorption in rain can also be determined from the radiative transfer approximation methods described in Section 7.1. The noise temperature

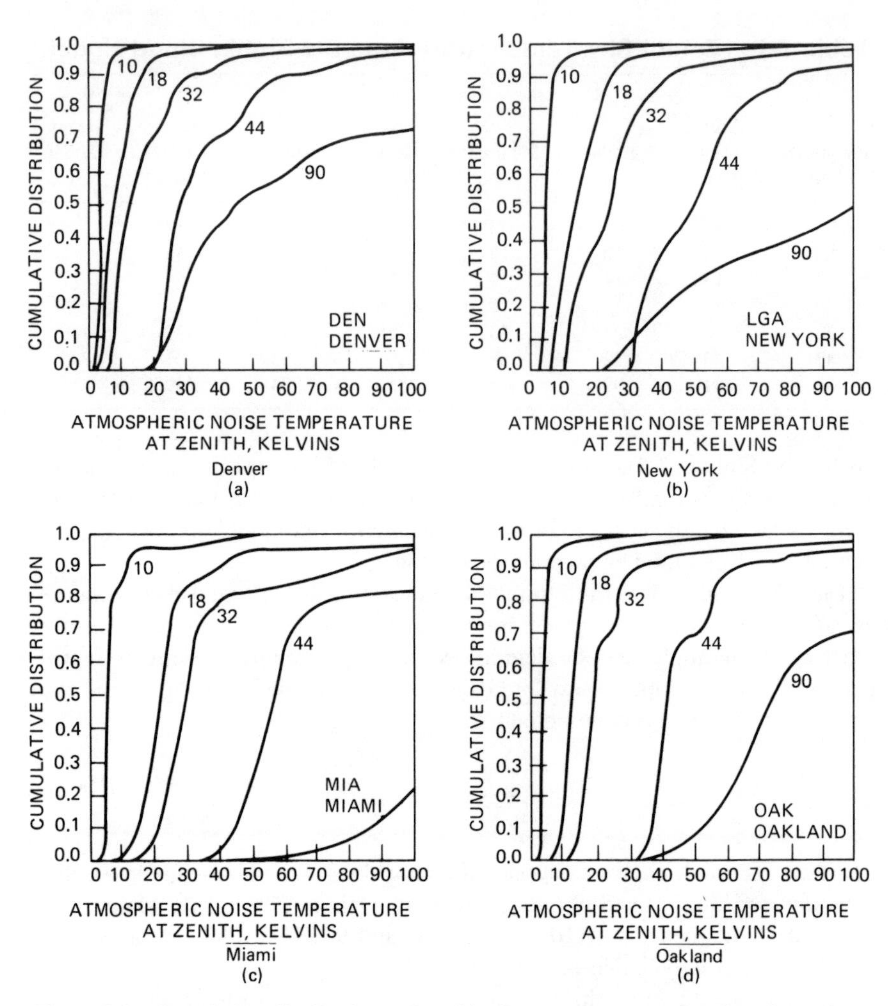

Figure 7-8. Cumulative distributions of zenith sky temperature at four locations, from the Slobin cloud model.

can be determined directly from the rain attenuation from [see Equation (7-5)]

$$T_s = T_m(1 - 10^{-[A(\text{dB})/10]}) \tag{7-7}$$

where T_m is the mean path temperature, in °K, and A(dB) is the total path rain attenuation, in dB. Note that the noise temperature is independent of frequency, i.e., for a given rain attenuation, the noise temperature produced will be the same, regardless of the frequency of transmission.

The noise is dependent on the path ambient temperature through the mean path temperature, T_m. Wolfsburg [7.5], using a model atmosphere, developed an expression for estimating the mean path temperature from the surface temperature,

$$T_m = 1.12T_g - 50 \tag{7-8}$$

where T_g is the surface temperature in °K. Using this expression, T_m ranges from about 255°K to 290°K for the surface temperature range of 0°C to 30°C.

The advent of satellite beacons in the late 1960s allowed for direct simultaneous measurements of rain attenuation and noise temperature on a slant path, and T_m could be determined directly. The best overall statistical correlation of the noise temperature and attenuation measurements occurs for T_m values between 270°K and 280°K for the vast majority of the reported measurements [7.6–7.8].

Figure 7-9 shows the noise temperature calculated from Equation (7-7) as a function of path attenuation for the range of values of T_m from 270°K to 280°K. The noise temperature approaches saturation, i.e., the value of T_m, fairly quickly above attenuation values of about 10 dB. Below that value the selection of T_m is not very critical. The center line (T_m = 275°K) serves as the best prediction

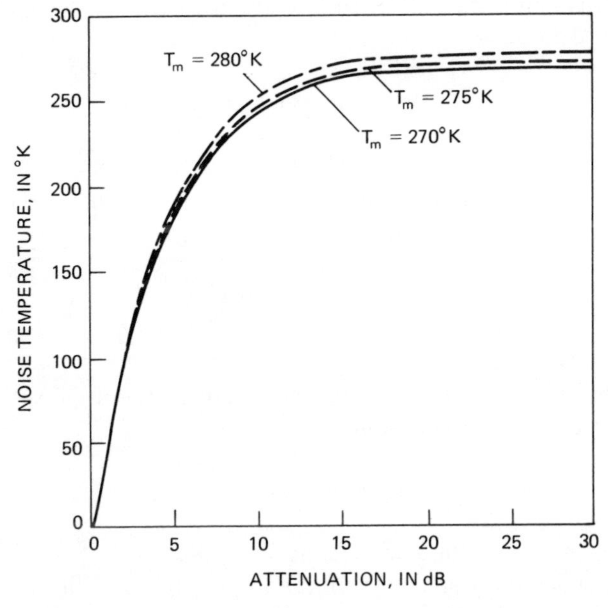

Figure 7-9. Noise temperature as a function of rain attenuation on a slant path.

curve for T_s. The noise temperature rises quickly with attenuation level. It is 56°K for a 1 dB fade, 137°K for a 3 dB fade, and 188°K for a 5 dB fade level.

The noise temperature introduced by rain will add directly to the receiver system noise figure, and will degrade the overall performance of the link. The *noise power increase* occurs coincident with the *signal power decrease* due to the rain fade; both effects are additive and contribute to the reduction in link carrier to noise ratio.

The increase in receiver system noise figure caused by noise from a rain fade is shown in Table 7-2 for systems with noise figures from 2 to 10 dB. The rain is seen to degrade the 2 and 3 dB systems significantly at all fade levels, producing effective noise figures of 4.01 and 4.67 dB respectively at a 30 dB fade level. This increase in system noise figure adds directly to the signal loss produced by the rain attenuation. For example, consider a 2 dB noise figure system experiencing a 15 dB rain fade. The total reduction in system carrier to noise ratio would be 15 + 1.96 or nearly 17 dB below that for nonrain conditions.

7.4. NOISE FROM SURFACE EMISSIONS

Energy incident to the surface of the Earth at some angle ϕ, of power p_i, will be partly transmitted and absorbed (p_t), partly specularly reflected (p_r), and, if the surface is rough, partly scattered (p_s). This process is shown graphically in Figure 7-10.

The reflective properties of the surface are quantitatively determined by the emissivity ϵ, defined as

$$\epsilon(\phi) = \frac{p_i - (p_r + p_s)}{p_i} \tag{7-9}$$

where the components of power are given above. For smooth surfaces, $p_s \ll p_r$, and the emissivity reduces to

$$\epsilon(\phi) = \frac{p_i - p_r}{p_i} \tag{7-10}$$

The brightness temperature (or apparent temperature) of the surface, at the angle of incidence ϕ, is found from

$$T_a(\phi) = \epsilon(\phi)\, T_{\text{surf}} \tag{7-11}$$

where T_{surf} is the temperature of the surface, in °K.

The differentiation of smooth surfaces from rough (or diffuse) surfaces is determined from the Rayleigh criterion,

Table 7-2. Increase in Receiver System Noise Figure Caused by Noise Contribution from a Rain Fade.

Rain Fade Level (dB)	Noise Contribution Due to Rain (°K)	''Effective'' Noise Figure for Given System Noise Figure (dB)				
		2 dB (170°K)	3 dB (290°K)	4 dB (438°K)	6 dB (865°K)	10 dB (2610°K)
1	56	2.50	3.41	4.32	6.21	10.08
3	137	3.12	3.92	4.74	6.48	10.20
5	188	3.47	4.21	4.98	6.65	10.27
10	247	3.85	4.53	5.25	6.83	10.35
15	266	3.96	4.62	5.33	6.89	10.37
20	272	3.99	4.65	5.35	6.91	10.38
30	274	4.01	4.67	5.37	6.91	10.39

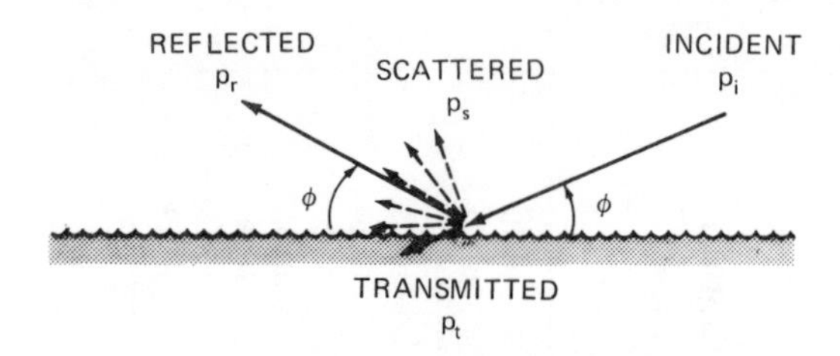

Figure 7-10. Components of energy incident on surface of the earth.

$$h = \frac{\lambda}{8 \cos \phi} \tag{7-12}$$

where h is the obsticle height, λ is the wavelength, and ϕ is the angle of incidence.

When $h > \lambda/8 \cos \phi$, the surface appears "rough" and more energy appears in the scattered component p_s. Under conditions of very rough surfaces, the scattered power predominates and the specular component is eliminated.

When $h \ll \lambda/8 \cos \phi$, the surface is considered smooth, and the specular reflection component predominates.

Figure 7-11 shows several examples of brightness temperatures determined from observed emissitivity for fresh water and sea water at frequencies from 1 to 50 GHz [7.9]. The curves for fresh water and sea water are indistinguishable for frequencies above 5 GHz, at either polarization.

Figure 7-12 shows a similar plot developed from emissitivity calculations for soil at 16.76 GHz as a function of incidence angle [7.2]. The brightness temperatures of land surfaces are higher than those of water surfaces because of lower dielectric constants. The effect of moisture in the soil lowers the brightness temperature, as seen in the figure.

The effect of soil surface roughness depends on the frequency. At frequencies near 1 GHz, deep vegetation such as dense woods provides a rough surface deep enough to eliminate the specular reflection from the surface. At 10 GHz, shrubbery will suffice, and at 35 GHz short grass is enough.

For the vast majority of practical space communications links, noise from surface emissions is not a major problem. The main antenna beam is not illuminated, and only energy which enters through the sidelobe pattern will affect the system. As an example, even if a worst case 290°K reflected emission were to enter the system through a sidelobe, which is typically at least 20 dB below the main lobe gain, the contribution to the system noise temperature would be 2.9°K or less. Careful site selection and reasonably well designed antenna and

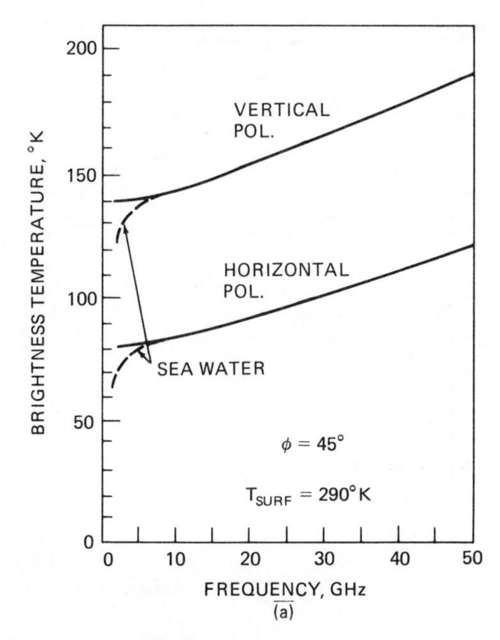

Figure 7-11. Brightness temperature of smooth fresh water and smooth sea water (dashed curves) as a function of frequency.

feed systems can protect from unwanted surface emissions due to both natural and man-made sources.

7.5. NOISE FROM EXTRATERRESTRIAL SOURCES

The Sun is a strong variable noise source with a noise temperature ranging from about 1,000,000°K at 100 MHz to 10,000°K at frequencies above 10 GHz, under quiet Sun conditions, with the Sun fully illuminating the antenna main beam (see Figure 7-3). Large increases will occur when the Sun is disturbed. Needless to say, direct viewing of the Sun by a satellite link will cause immediate problems.

Below about 2 GHz, galactic noise will be a factor, while above 2 GHz only cosmic background noise (2.7°K will be present (see Figure 7-3).

Figure 7-13 shows a plot of the radio sky at 250 MHz in equatorial coordinates (declination versus right ascension) [7.10]. A geostationary satellite as seen from the Earth appears as a horizontal line of fixed declination between +8.7° and −8.7°, shown by the shaded band on the figure.

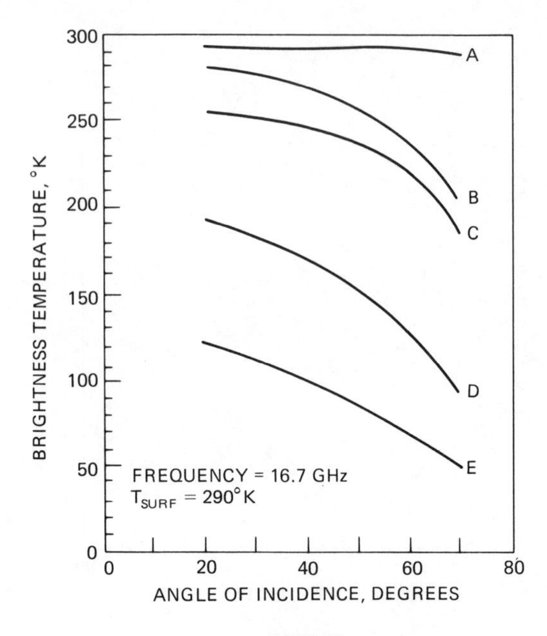

Figure 7-12. Brightness temperature of soil at 16.7 GHz, as a function of moisture content and incidence angle.

The contours of Figure 7-13 are in units of 6°K above 80°K, the values corresponding to the coldest parts of the sky, For example, at 1800 h and 0° declination, the countour value is 37. The brightness temperature at 250 MHz is then $6 \times 37 + 80 = 302$°K. The brightness temperature for another frequency, f_i, is found from

$$T_b(f_i) = T_b(f_0)\left(\frac{f_i}{f_0}\right)^{-2.75} + 2.7 \tag{7-13}$$

For example, at a frequency of 1 GHz, the above value becomes

$$T_b(1 \text{ GHz}) = 302\left(\frac{1}{0.25}\right)^{-2.75} + 2.7 = 9.4°\text{K}$$

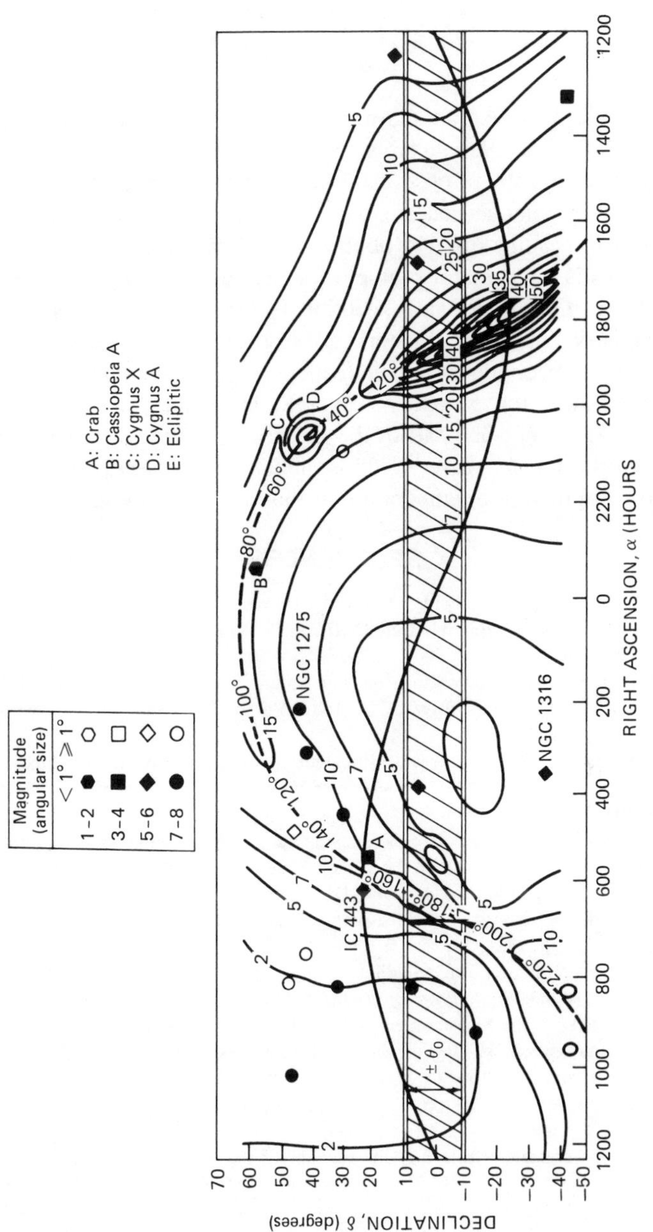

Figure 7-13. The radio sky at 250 MHz in the region around the geostationary satellite orbital arc.

and at 4 GHz

$$T_b(4\text{ GHz}) = 302\left(\frac{4}{0.25}\right)^{-2.75} + 2.7 = 2.8°\text{K}$$

Several relatively strong nonthermal sources, such as Cassiopeia A, Cygnus A and X, and the Crab Nebula are shown on the figure (marked A through D). They are not in the zone of observation for geostationary satellites, and would only be in view for nongeostationary orbits for a very small segment of time.

Except for direct line-of-sight observation of the Sun, noise from extraterrestrial sources is not a major problem for the vast majority of space communications links. One important exception is the NASA deep space network (DSN) telemetry links which operate at very low system noise temperatures. The DSN station receiving system equivalent noise temperature for the *Voyager* deep space mission, for example, is 28.5°K, at an operating frequency of 8.4 GHz [7.11]. At the noisiest part of the galactic sky, near the galactic nucleus, the noise temperature increase would be about 2.8°K, resulting in an increase of about 10% in the noise temperature of the receiver system.

REFERENCES

7.1. CCIR, Report 670, "Worldwide Minimum External Noise Levels, 0.1 Hz to 100 GHz," Recommendations and Reports of the CCIR, 1978, Vol. 1, *Spectrum Utilization and Monitoring*, Geneva, 1978.

7.2. CCIR, Report 720-1, "Radio Emission from Natural Sources Above About 50 MHz," Recommendations and Reports of the CCIR, 1982, Vol. 5, *Propagation in Non-Ionized Media*, pp. 151–166, Geneva, 1982.

7.3. Smith, Ernest K., "Centimeter and Millimeter Wave Attenuation and Brightness Temperature Due to Atmospheric Oxygen and Water Vapor," *Radio Science*, Vol. 17, No. 6, pp. 1455–1464, Nov.-Dec. 1982.

7.4. Slobin, S. D., "Microwave Noise Temperature and Attenuation of Clouds: Statistics of These Effects at Various Sites in the United States, Alaska, and Hawaii," *Radio Science*, Vol. 17, NO. 6, pp. 1443–1454, Nov. -Dec. 1982.

7.5. Wolfsburg, K. N., "Sky Noise Measurements at Millimeter Wavelengths," *Proc. of the IEEE*, Vol. 52, pp. 321–322, March 1964.

7.6. Ippolito, L. J., "Effects of Precipitation on 15.3 and 31.65 GHz Earth–Space Transmissions with the ATS-V Satellite," *Proc. of the IEEE*, Vol. 59, pp. 189–205, Feb. 1971.

7.7. Strickland, J. I., "The Measurement of Slant Path Attenuation Using Radar, Radiometers and a Satellite Beacon," *J. Res. Atmos.*, Vol. 8, pp. 347–358, 1974.

7.8. Hogg, D. C., and Chu, Ta-shing, "The Role of Rain in Satellite Communications," *Proc. of the IEEE*, Vol. 63, No. 9, pp. 1308–1331, Sept. 1975.

7.9. Wilheit, T. T., and Chang, A. T. C., "An Algorithm for Retrieval of Ocean Surface and Atmospheric Parameters from the Observations of the Scanning Multichannel Microwave Radiometers (SMMR)," NASA Technical Memorandum 80277, Goddard Space Flight Center, Greenbelt, MD, May 1979.

7.10. Ko, H. C., and Kraus, J. D., "A Radio Map of the Sky at 1.2 Meters," *Sky and Telescope*, Vol. 16, pp. 160–161, February 1957.

7.11. Yuen, Joseph H., editor, "Deep Space Telecommunications Systems Engineering," NASA JPL Publication 82-76, July 1982.

CHAPTER 8
SCINTILLATION, BANDWIDTH COHERENCE, AND OTHER PROPAGATION FACTORS

In this chapter, several propagation factors introduced in Section 2.6 which have not yet been fully discussed in previous chapters are described. The factors described here are generally not as severe in earth–space communications as those described in earlier chapters, and they can be considered as secondary effects for most typical space communications links. There are special situations, however, where the effects can become significant, and these situations will be pointed out.

Subjects covered here include: scintillation, bandwidth coherence, angle of arrival variations, and antenna gain degradation.

8.1. SCINTILLATION

Scintillation describes the condition of rapid fluctuations of the signal parameters of a radiowave caused by time dependent irregularities in the transmission path. Signal parameters affected include amplitude, phase, angle of arrival, and polarization.

Scintillation effects can be produced in both the ionosphere and in the troposphere. Electron density irregularities occuring in the ionosphere can affect frequencies up to about 6 GHz, while refractive index irregularities occurring in the troposphere cause scintillation effects in the frequency bands above about 3 GHz.

The mechanisms and characteristics of ionospheric and tropospheric scintillation differ, and they are discussed separately in the following sections.

8.1.1. Ionospheric Scintillation

Ionospheric scintillation is produced by electron density fluctuations near the altitude of maximum electron density, the F region, at approximately 200–400 km in altitude. These conditions are most prevalent in the equatorial regions, at high latitude locations, and during periods of high sunspot activity. Ionospheric scintillations have been observed at frequencies from 20 MHz through 6 GHz, with the bulk of data being amplitude scintillation observations in the VHF (30–300 MHz) bands [8.1]. Scintillations can be very severe in the frequency bands below 300 MHz, and often are the limiting factor for reliable communications system performance in the VHF bands.

Figure 8-1 shows an example of amplitude and phase scintillations observed on a satellite path, at frequencies of 150 and 400 MHz [8.2]. The measurements involved coherent signals from a U.S. Navy Navigational System Satellite and a 25.6 meter receiving antenna in Massachusetts. The figure shows a one minute segment of signal amplitude and differential phase. The 400 MHz data show peak-to-peak amplitude fluctuations in excess of 10 dB, while the 150 MHz values exceed 30 dB. The differential phase data show both large scale size (slow) and small scale size (rapid) fluctuations.

Several measures or indices are available to quantitatively classify scintillation effects. One of the most useful is the *S* index, defined as follows [8.3]:

$$S_4 = \frac{\sigma_x}{m_x} \tag{8-1}$$

where σ_x is the standard deviation of received power, and m_x is the mean.

Another often used measure is the *SI* index, defined as [8.4]

$$SI = \frac{P_{\max} - P_{\min}}{P_{\max} + P_{\min}} \tag{8-2}$$

where the P's represent power. The third peak down from the maximum peak and the third minimum up from the absolute minimum are used, to avoid overemphasizing extreme conditions.

The standard deviaton or variance of the logarithm of the received power, σ_x and σ_x^2 respectively, in dB, are also used to characterize to intensity of scintillation.

For the sample data segment shown in Figure 8-1, the *S* values are 0.28 and 1.00 at 400 and 150 MHz, the *SI* values are 0.695 and 0.998, respectively, and the σ_x values are 1.27 dB and 5.50 dB, respectively.

Ionospheric scintillation tends to be the most severe in equatorial, auroral, and polar latitudes, with periods of increased magnetic activity. Table 8-1 shows the percentage of occurrence of scintillations for various locations and frequencies as compiled by the CCIR [8.1]. Section (a) of the table is for equatorial latitudes with scintillations > 10 dB peak-to-peak; section (b) is for auroral and sub-auroral latitudes with scintillations > 12 dB at 137 MHz; and section (c)

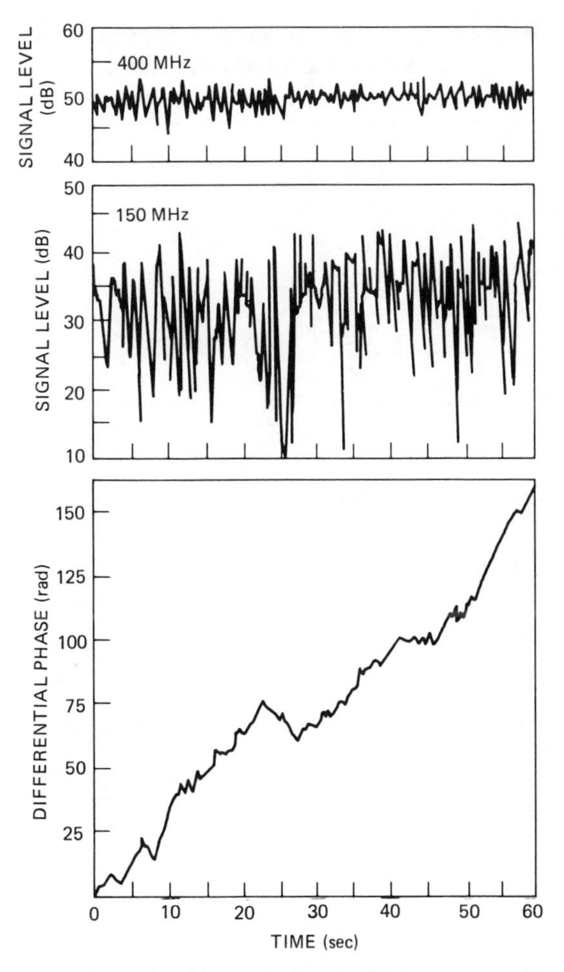

Figure 8-1. Example of ionospheric scintillation on a satellite path.

is for auroral latitudes and > 10 dB at 254 MHz. The planetary magnetic index, K_p listed in the table is a measure of magnetic activity. The table shows how scintillation activity increases with magnetic activity at auroral and sub-auroral latitudes.

Equatorial scintillation is often characterized by a sudden onset, and its occurrence is heavily dependent on location within the equatorial region.

Mid-latitude locations show a well established maximum near midnight, with a second maximum around midday at some locations. These occurrences are most pronounced during summer periods.

Scintillation levels increase considerably in the high latitude auroral regions.

Table 8-1. Percentage of Occurence of Ionospheric Scintillation for Several Global Locations.

(a) ⩾ 10 dB peak to peak, equatorial latitudes

Site	Frequency	Day	Night
Huancayo (Peru)	137 MHz	(0400–1600 LT)	(1600–0400 LT)
	254 MHz	3%	14%
		2	7
Accra (Ghana)	137 MHz	(0600–1800 LT)	(1800–0600 LT)
		0.4	14

(b) ⩾ 12 dB peak to peak at 137 MHz, sub-auroral and auroral latitudes

Site	K_p	Day (0500–1700 LT)	Night (1700–0500 LT)
Sagamore Hill (Massachusetts)	0 to 3+	0	1.4
	> 3+	0.1	2
Goose Bay (Labrador)	0 to 3+	0.1	1.8
	> 3+	1.6	6.8
Narssarssuaq (Greenland)	0 to 3+	2.9	18
	> 3+	19	45

(c) ⩾ 10 dB peak to peak at 254 MHz, auroral latitudes

Site	K_p	Day (0600–1800 LT)	Night (1800–0600 LT)
Goose Bay (Labrador)	0 to 3+	0.1	0.1
	> 3+	0.3	1.2
Narssarssuaq (Greenland)	0 to 3+	0.1	0.9
	> 3+	2.6	8.4

LT = local time.
SOURCE: CCIR [8.1].

At polar latitudes, the scintillation level is high but still somewhat lower than at the auroral latitudes.

Table 8-2 summarizes the general characteristics of the ionospheric scintillation parameters for equatorial, mid-latitude, and high latitude regions. Further details of the characteristics of the morphology of ionospheric scintillation are provided in CCIR Report 263-4 [8.1].

Ionospheric scintillations, although primarily a low frequency effect, have been observed on microwave frequency links as well. Equatorial scintillations have been measured on INTELSAT satellite links at frequencies as high as 6 GHz [8.5]. Peak-to-peak fluctuations exceeded 4.8 dB for 0.01% of the time at 6 GHz in Hong Kong, and exceeded 2.8 dB for the same conditions at Guam.

Table 8-2. General Characteristics of Ionospheric Scintillation.

	Region		
Parameter	Equatorial	Mid-latitude	High-latitude
Activity level	Exhibits greatest extremes	Generally very quiet to moderately active	Generally moderately active to very active
Diurnal	Maximum-night-time Minimum-day-time	Maximum-night-time Sporadic-day-time	Maximum-night-time Minimum-day-time
Seasonal	Longitudinal dependent Peaks in equinoxes Accra, Ghana Maximum-November and March Minimum-solstices Huancayo, Peru Maximum-October through March Minimum-May through July Kwajalein Islands Maximum-May Minimum-November and December	Maximum-spring Minimum-winter	Maximum-summer Minimum-winter
Solar cycle	Occurence increases with sunspot number	Insufficient data	Occurrence increases with sunspot number

Table 8-2. Continued

	Region		
Parameter	Equatorial	Mid-latitude	High-latitude
Magnetic activity	Longitudinal dependent Accra, Ghana Occurrence decreases with K_p Huancayo, Peru March equinox - Occurrence decreases with K_p June solstice - Occurrence increases with K_p September equinox. 0000–0400 h (local time) Occurrence increases with K_p	Independent of K_p	Occurrence increases with K_p

SOURCE: CCIR [8.1].

Amplitude scintillations have been monitored at equatorial sites using the 1.54 GHz MARISAT satellite [8.6], [8.7]. The maximum scintillation observed at Hauncayo, Peru, for a 20 month period was 8 dB peak to peak. The scintillation increased during periods of high sunspot activity to greater than 27 dB peak to peak as Ascension Island, and 7 to 9 dB peak to peak at Huancayo and Natal, Brazil.

The frequency dependence of ionospheric scintillation was studied on a satellite link by employing ten simultaneously transmitted signals ranging from 137 to 2891 MHz [8.8]. Measurements of scintillation at equatorial and auroral locations (Ancon, Peru; Kwajalein Island; Fairbanks, Alaska) showed a $1/f^{1.5}$ frequency dependence for moderate ($S_4 < 0.4$) amplitude scintillations, and a $1/f$ dependence for phase scintillations. Earlier measurements using galactic sources and more limited frequency ranges have shown $1/f$ frequency dependence coefficients ranging from about 1.2 to 2.1 for amplitude scintillations [8.2].

Frequency dependence measurements are subject to a wide range of measurement errors, however, and a consistent analytically based method of frequency extrapolation for ionospheric scintillation is not presently available.

Attempts to analytically describe and theoretically predict ionospheric scintillation have met with limited success. A complete wave propagation model for a volume of random refractive index irregularities along the path is not presently available; however, approximate solutions have been developed that are valid under specialized conditions. Traditional perturbation series solutions, such as the Born, Rytov, and Markov approximations, have been employed, as well as a thin phase diffraction screen model [8.9]. Table 8-3 summarizes the conditions, as described by Crane [8.2], under which the approximation methods can be applied to the evaluation of ionospheric scintillation effects on an earth–space path.

8.1.2. Tropospheric Scintillation

Tropospheric scintillation is typically produced by refractive index fluctuations in the first few kilometers of altitude and is caused by high humidity gradients and temperature inversion layers. The effects are seasonally dependent, vary day-to-day, and vary with the local climate. Tropospheric scintillation has been observed on line-of-site links up through 10 GHz and on earth–space paths at frequencies to above 30 GHz.

To a first approximation, the refractive index structure in the troposphere can be considered horizontally stratified, and variations appear as thin layers which change with altitude. Slant paths at low elevation angles, that is, highly oblique to the layer structure, thus tend to be affected most significantly by scintillation conditions.

Table 8-3. Approximation Methods for Ionospheric Scintillation Models.

	Method			
Condition	Born Approximation	Rytov Approximation	Single Thin Phase Screen	Markov Approximation
Weak scintillation produced by a thin region or a single dominant irregularity	✓	✓	✓	✓
Weak scintillation and a thick irregularity region	✓	✓		✓
Strong scintillation and a thin layer			✓	✓
Strong scintillation and a thick layer				✓

The general properties of the refractive index of the troposphere are well known. The index of refraction, n at radiowave frequencies is a function of temperature, pressure, and water vapor content. For convenience, since n is very close to 1, the refractive index properties are usually defined in terms of N units, or refractivity, as

$$N = (n - 1) \times 10^6 = \frac{77.6}{T}\left(p + 4810\,\frac{e}{T}\right) \tag{8-3}$$

where p is the atmospheric pressure in millibars (mb), e is the water vapor pressure in mb, and T is the temperature in degrees Kelvin. This expression is accurate to within 0.5% for frequencies up to 100 GHz.

Figure 8-2 shows the variation of N with altitude h for a mid-latitude U.S. Standard Atmosphere, 1966. The dashed curve on the figure shows an exponential model for N of the form

$$N = 315e^{-(h/7.36)} \tag{8-4}$$

recommended by the CCIR for use in the lower few kilometers, where radiowave effects predominate [8.10].

Small scale variations of refractivity, such as those caused by temperature inversions or turbulence, will produce scintillation effects on a radiowave.

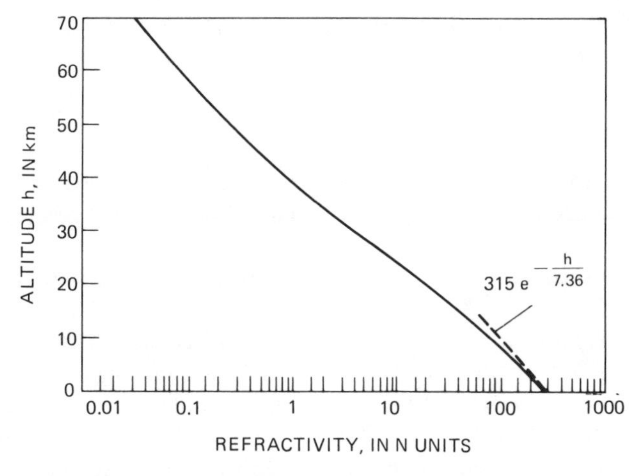

Figure 8-2. Refractivity profile for a mid-latitude dry U.S. standard atmosphere.

Quantitative estimates of the level of amplitude scintillation produced by a turbulent layer in the troposphere are determined by assuming small fluctuations on a thin turbulent layer and applying turbulence theory considerations [8.11]. Amplitude scintillation is expressed as

$$x(\text{dB}) = \log r \tag{8-5}$$

the log of the received power. The variance of the log of the received power, σ_x^2, is then found as

$$\sigma_x^2 = 42.25 \left(\frac{2\pi}{\lambda}\right)^{7/6} \int_0^L C_m^2(x)\, x^{5/6}\, dx \tag{8-6}$$

where $C_m(x)$ is a refractive index structure constant, λ is the wavelength, x is the distance along the path, and L is the total path length. A precise knowledge of the amplitude scintillation depends on $C_m(x)$, which is not easily available.

The results do give insight into the frequency dependence of amplitude scintillation, however. Equation (8-6) shows that the r.m.s. amplitude fluctuation, σ_x, varies as $f^{7/12}$. Measurements at 10 GHz which show a range of σ_x from 0.1 to 1 dB, for example, would scale at 100 GHz to a range of about 0.38 to 3.8 dB.

A tropospheric scintillation model based on measurements made in the northeastern United States is available from the CCIR [8.12]. A thin turbulent layer at an average height of 1 km is assumed, and empirical fits to satellite based measurements in Massachussetts at 7.3 GHz and 400 Mhz are used to determine the coefficients required. The resulting prediction equation for the root-mean-square (r.m.s.) amplitude fluctuations is

$$\sigma_x(\text{dB}) = (2.5 \times 10^{-2})\, f^{7/12}\, (\csc\theta)^{0.85}\, [G(D)]^{1/2} \qquad (8\text{-}7)$$

where f is the frequency, in GHz, θ is the elevation angle, in degrees, and $G(D)$ is an antenna aperture averaging factor determined from the antenna diameter D.

Figure 8-3 shows a plot of the predicted r.m.s. amplitude fluctuations σ_x for frequencies from 1 to 100 GHz as a function of elevation angle. A 10 meter diameter antenna with an efficiency of 56% was assumed. The r.m.s. fluctuations are seen to be quite low, particularly for elevation angles above 20°.

The model has shown good agreement with several sets of experimental measurements made in the U.S. and Japan, at frequencies from 400 MHz to 30 GHz. Appendix F presents the complete step by step prediction procedure for the CCIR Tropospheric Scintillation Model, including an example of its application.

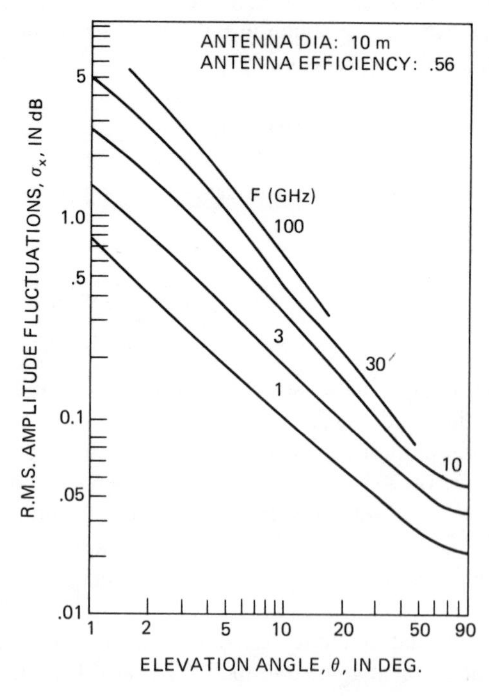

Figure 8-3. Amplitude scintillations predicted from the CCIR tropospheric scintillation model.

8.1.3. Amplitude Scintillation Measurements on Earth–Space Paths

The most predominant form of scintillation observed on earth–space communications links involves the amplitude of the transmitted signal, usually on the downlink. Since the frequencies of most of the reported events are above 2 GHz, the scintillation effects observed are most likely of tropospheric origin.

Several organizations reported scintillation measurements using the ATS-6 satellite at frequencies of 2, 20, and 30 GHz, and other satellite transmissions at 4, 6, and 7 GHz [8.13]. The measurements showed broad agreement for scintillations at high elevation angles (20–30°). In temperate climates the scintillations were on the order of 1 dB peak to peak in clear sky in the summer, 0.2–0.3 dB in winter, and 2–6 dB in cloud conditions. The rate of the scintillations varied over a large range, however, with fluctuations from 0.5 Hz to over 10 Hz. A much slower fluctuation component, with a period of 1–3 minutes, was often observed along with the more rapid scintillations discussed above. At low elevation angles (below 10 degrees) scintillation effects increased drastically. Deep fluctuations of 20 dB or more were observed, with durations of a few seconds in extent.

Figure 8-4 shows an example of amplitude scintillation measurements at 2 and 30 GHz made with the ATS-6 at Columbus, Ohio [8-14]. The elevation angles were 4.95° (a), and 0.38° (b). Measurements of this type were made in clear weather conditions up to an elevation angle of 44°, and the data are summarized in Figure 8-5, where the mean amplitude variance is plotted as a function of elevation angle. The curves on the figure represent the minimum r.m.s. error fits to the assumed cosecant power law relation

$$\sigma_x^2 = A(\csc\theta)^B \tag{8-8}$$

where θ is the elevation angle. The resulting B coefficients, as shown on the figure, compare well within their range of error with the expected theoretical value of 1.833 for a Kolmogorov-type turbulent atmosphere, and 1.7 for the CCIR scintillation model [Equation (8-7)].

Similar measurements were taken at 19 GHz with the COMSTAR satellites at Holmdel, N.J. [8.15]. Both horizontal and vertical polarized signals were monitored, at elevation angles from 1 to 10 degrees. Amplitude scintillations at the two polarizations were found to be highly correlated, leading the authors to conclude that the scintillations were independent of polarization sense.

Table 8-4 summarizes the results of low elevation angle amplitude scintillation measurements observed on satellite links in North America, Europe, and Japan [8.13]. The measurements, all observed under clear sky conditions, show how the severity of scintillation increases dramatically for elevation angles of about 3 degrees or lower.

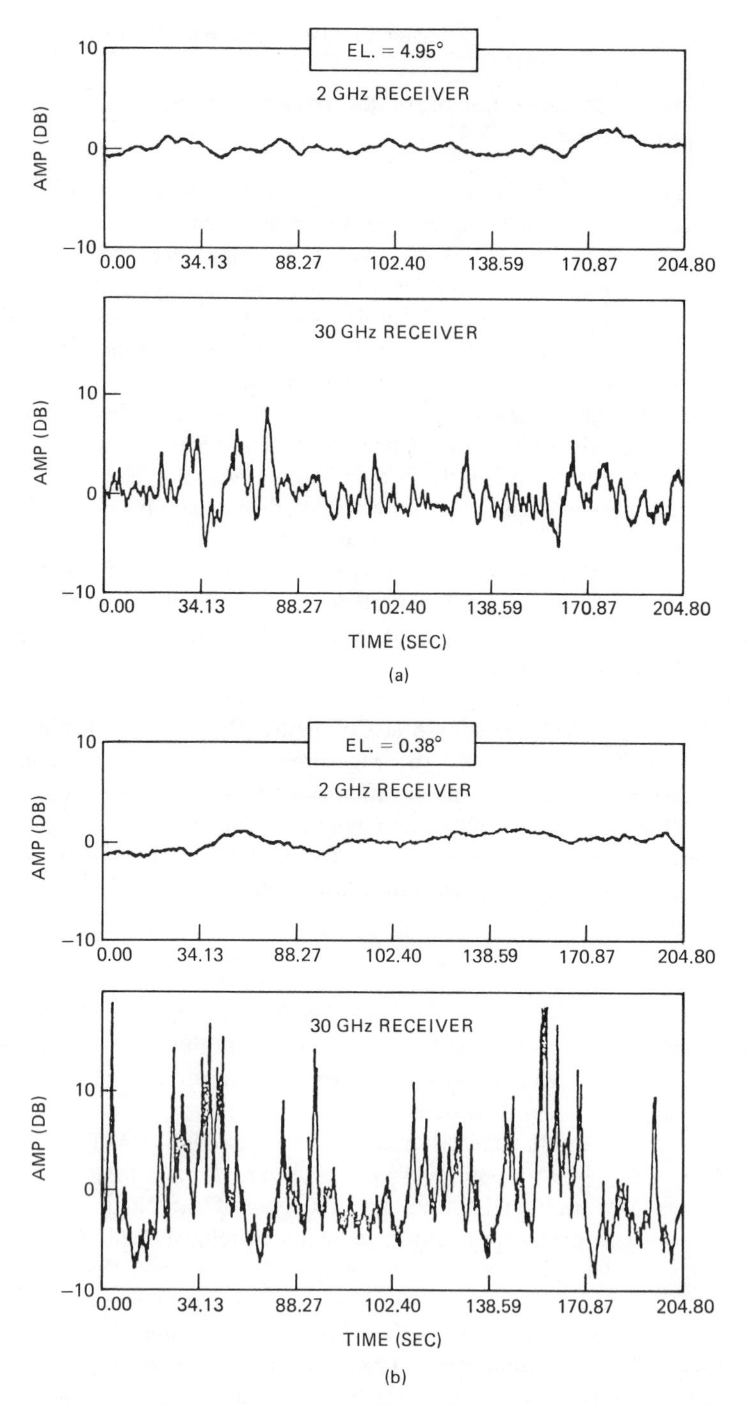

Figure 8-4. Amplitude scintillations on a satellite link at low elevation angles.

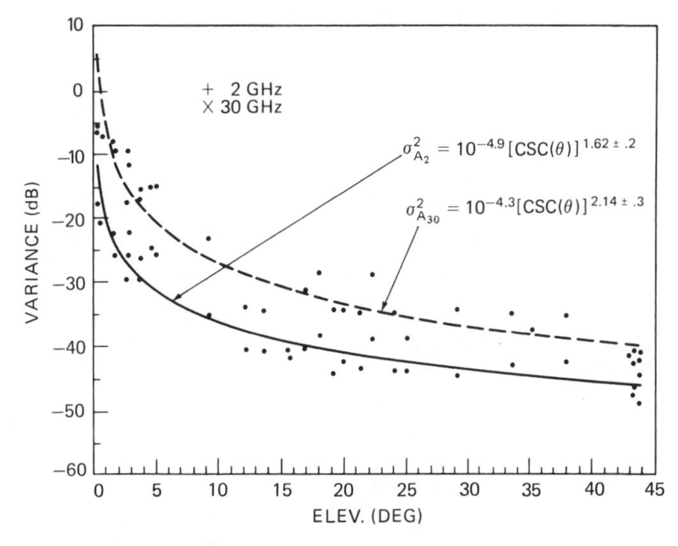

Figure 8-5. Mean amplitude variance for clear weather conditions, at 2 and 30 GHz, as a function of elevation angle.

Amplitude scintillations have been observed on satellite links when clouds are present on the path. Figure 8-6 shows an example of scintillations produced on a 20 GHz link in North Carolina as a cumulus cloud passed through the satellite beam [8.16]. The amplitude fluctuations increased from about 0.5 dB to 3 dB peak to peak during the passage of the cloud. The fluctuation rate was fairly constant at about 16 cycles per minute (0.27 Hz). The scintillations lasted for about 200 seconds, then settled back to the initial levels. The elevation angle to the satellite for this observation was 47 degrees.

Similar results for cloud induced scintillations on 19 and 28 GHz links at Crawford Hill, NJ, have also been reported [8.17]. Over 1000 events at 28 GHz were observed in two summer months. The mean duration of the scintillation events was short, i.e., abut 1.3 seconds for fades greater than 1 dB peak to peak. The fluctuation rate averaged about 0.3 Hz, and the scintillations were found to be independent of signal polarization. Instantaneous scintillation amplitudes at 19 and 28 GHz along identical paths were highly correlated, with the intensity at 28 GHz 1.2 times that at 19 GHz, in dB. This result is consistent with the $f^{7/12}$ frequency dependence predicted by turbulence theory, as described in the previous section.

8.2 BANDWIDTH COHERENCE

The question of whether the earth's atmosphere will produce a limitation on the information bandwidth (or, equivalently, the data rate) of a space telecommu-

Table 8-4. Low Elevation Angle Amplitude Scintillation Statistics Observed on Satellite Links.

Satellite and Frequency	Location of Measurements	Antenna Diameter	Elevation Angle	Period of Measurement	Peak-to-peak Amplitude Exceeded for Given Percentages of: Whole Period 0.1%	Whole Period 0.3%	Whole Period 1%	Worst Month 0.1%	Worst Month 1%
Anik-A, 4 and 6 GHz	Eureka, Canada	4.6 m	1°		—	—	—	20 dB	11 dB
Intelsat-IV, 4 and 6 GHz	Goonhilly, UK		6.5°			3 dB		4 dB	3 dB
Intelsat-IV, 4 and 6 GHz combined	Goonhilly, UK		6.5°						
	Bahrain	27.5 m		9 months				6 dB	5.2 dB
	Yamaguchi, Japan		9°						
7 GHz	Ottawa, Canada	9.1 m	1°		—	—	15 dB	—	—
			2–3°	Summer period	—	—	8 dB	—	—
			5°		5.5 dB	—	2.8 dB	—	—
OTS, 11.8 GHz	Svalbard, Norway	3 m	3.2°	April-December 1979–June-August 1980	—	—	—	6.5 dB	3.4 dB

SOURCE: CCIR Report 564-2 [8.2].

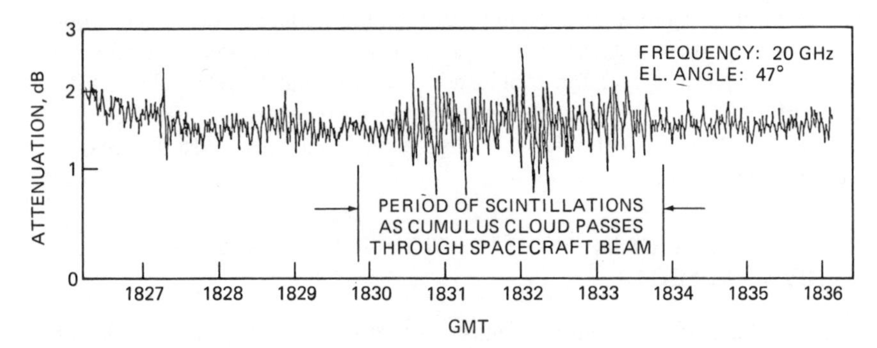

Figure 8-6. Amplitude scintillations produced by a cloud passing through the satellite path.

nications link is important, particularly as space communications systems move to wider bandwidths at higher carrier frequencies and digital techniques involving large channel bandwidths are employed. Wideband signals are sensitive to amplitude and phase dispersion effects over the entire bandwidth, and if the dispersions are large, coherency across the bandwidth would be lost and the information on the signal could be distorted or lost. Present satellite links operate with individual transponder bandwidths on the order of 100 MHz or less, but bandwidths of 500 MHz and higher are being considered for future systems.

Theoretical studies of propagation effects on wide bandwidth transmissions have indicated that the atmosphere can support several gigahertz of bandwidth at carrier frequencies above 10 GHz [8.18, 8.19]. The results show that bulk attenuation effects will predominate over any bandwidth reduction effect such as frequency selective fading or severe amplitude and/or phase dispersion.

For example, at carrier frequencies above 10 GHz, and for very severe rain conditions, rain attenuations of over 100 dB would be observed before frequency dispersion would become appreciable. For storms in temperate climates, *bandwidths* in excess of 3.6 GHz must be used before bandwidth distortion becomes important [8.19].

Experimental verification of bandwidth coherency on an earth–space path was first demonstrated, at 20 and 30 GHz, with the ATS-6 satellite [8.16]. Nine sidebands spread out over a 1.44 GHz bandwidth were used to determine the frequency and phase characteristics of the atmosphere during rain. The results showed no measurable selective fading over the 1.44 GHz bandwidth, even for severe rain fades. Amplitude and phase variations were within the measurement accuracy of the receiving equipment, i.e., ± 0.5 dB and 4 degrees, respectively.

Similar results were reported on measurements with the COMSTAR satellites [8.20]. In this experiment, amplitude and phase dispersion over a 528 MHz bandwidth at 28 GHz were measured for a one year period, and no evidence of any amplitude or phase dispersion other than the frequency dependence due to the bulk properties of water in rain was observed. The authors conclude: "amplitude and phase dispersion should not pose a problem for wideband (on the order of 1 GHz) earth–space communications systems operating at frequencies greater than 10 GHz with relatively large elevation angles (>15 degrees) from the earth terminals."

8.3. ANTENNA GAIN DEGRADATION AND ANGLE OF ARRIVAL EFFECTS

The antenna gain in a communication system is generally defined in terms of the antenna behavior when illuminated by a uniform plane wave. Amplitude and phase fluctuations induced by the atmosphere, however, can produce perturbations across the antenna aperture, resulting in a reduction of total power available at the antenna feed. The resulting effect on the antenna will look to the system like a loss of antenna gain, or a *gain degradation*. This loss may be interpreted in two ways: (1) in terms of angle of arrival fluctuations which cause the incident signal to arrive from different directions other than the maximum gain direction, or (2) as a phase dispersion of the ray paths which reflect from the surface of the antenna reflector, resulting in a summation at the feed point of rays which no longer are in phase.

Gain degradation will increase as the electrical receiving aperture size increases, hence the problem will become more intense as the operating frequency and/or the physical aperture size increase. Also, the effect will be more pronounced as the path length in the atmosphere increases, i.e., for low elevation angles.

Estimates of gain degradation have been determined by application of conventional atmospheric turbulence theory to an earth–space path [8.21]. Figure 8-7 presents gain degradation as a function of elevation angle at a frequency of 30 GHz, and for antenna beamwidths from 0.3 to 0.05 degrees. Gain degradation is less than 2 dB for elevation angles above 10 degrees, even for 0.05 degree beamwidth.

Figure 8-8 shows the frequency and antenna beamwidth dependence of the gain degradation, as a function of elevation angle. There is very little dependence on frequency, and very little gain degradation except for elevation angles well below 10 degrees. These curves represent gain degradation due to atmospheric turbulence only, and additional effects could be expected from atmospheric gasses, clouds, and rain in the beam.

Angle of arrival measurements in the presence of rain were made with the 28 GHz COMSTAR beacon [8.22]. The measurements were made at an elevation angle of 38 degrees, and with two receiving apertures , one at 7 meters and the other at 0.6 meters. The beamwidths were 0.1 degrees and 1×1.4 degrees,

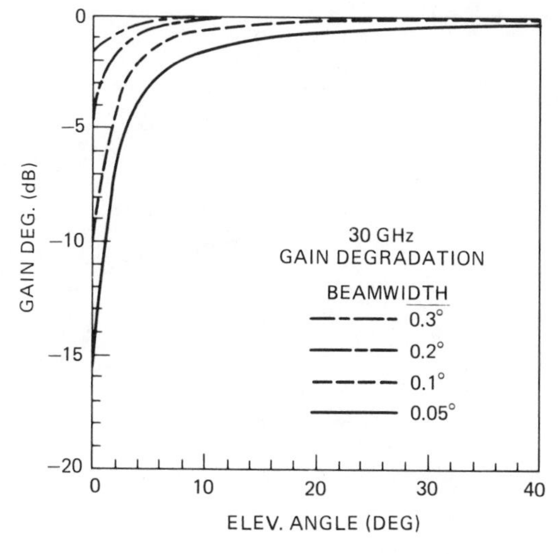

Figure 8-7. Gain degradation at 30 GHz due to atmospheric turbulence.

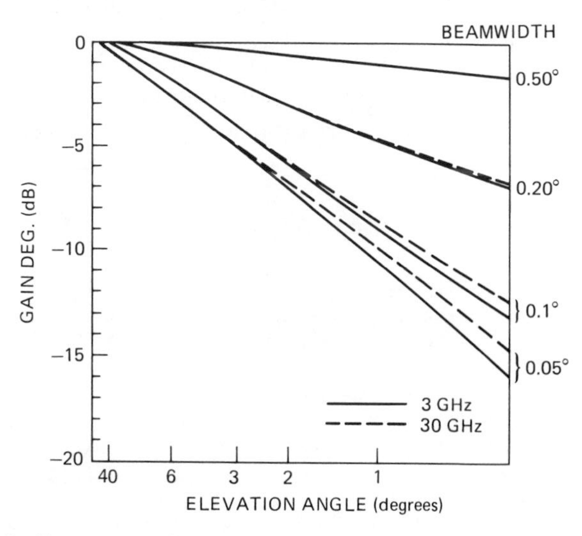

Figure 8-8. Frequency and antenna beamwidth dependence of gain degradation.

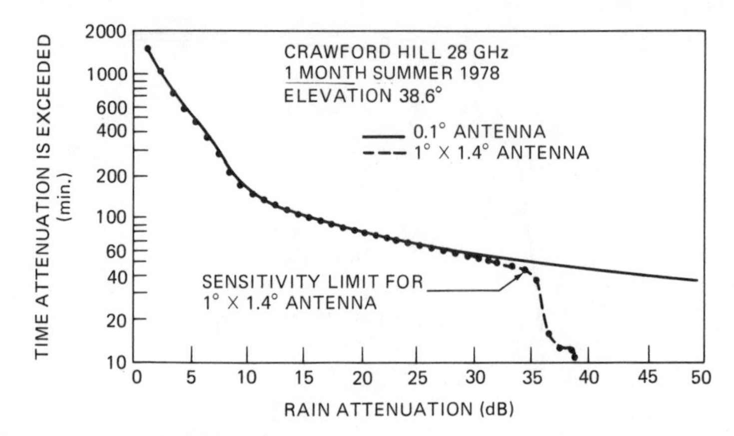

Figure 8-9. Cumulative distribution of rain attenuation for angle of arrival measurements.

respectively. Very little difference in the cumulative distributions of rain attenuation measured with the two antennas was observed, as seen in Figure 8-9. Above about 28 dB rain attenuation, the 1 × 1.4 degree antenna distribution becomes dominated by the noise level of the receiver. The measured rain attenuation is essentially independent of antenna beamwidth for attenuations up to about 30 dB. The spread was less than ±0.5 dB for 98% of the time up to an attenuation of 10 dB and less than ±1.0 dB up to almost 20 dB. This translates to peak angle of arrival fluctuations of less than 0.03 degrees and 0.04 degrees, respectively.

In summary, the available analyses and direct measurements indicate that antenna gain degradation and angle of arrival effects should not be a major factor for typical earth–space communications links operating at elevation angles well above 10 degrees and at frequencies up to 30 GHz.

REFERENCES

8.1. CCIR, Report 263-4, "Ionospheric Effects upon Earth–Space Propagation," in Volume VI, *Propagaton in Ionized Media*, Recommendations and Reports of the CCIR—1978, International Telecommunications Union, Geneva, pp. 71–89, 1978.

8.2. Crane, R. K., "Ionospheric Scintillation," *Proc. of the IEEE*, Vol. 65, No. 2, pp. 180–199, Feb. 1977.

8.3. Briggs, B. H., and Parkin, I. A., "On the Variation of Radio Star and Satellite Scintillations with Zenith Angle," *J. Atmos. Terr. Phys.*, Vol. 25, pp. 339–365, June 1963.

8.4. Whitney, H. E., Aarons, J., and Malik, C., "A Proposed Index for Measuring Ionospheric Scintillation," *Planet. Space Sci.*, Vol. 7, pp. 1069–1073, 1969.

8.5. Taur, R. R., "Ionospheric Scintillation at 4 and 6 GHz," *COMSAT Technical Review*, Vol. 3, pp. 145–163, Spring 1973.

8.6. Basu, S., Basu, D., Mullen, J. P., and Bushby, A., "Long Tern 1.5 GHZ Amplitude Scintillation Measurements at the Magnetic Equator," *Geophys. Res. Lett.*, Vol. 7, pp. 259-262, April 1980.

8.7. Aarons, J., Whitney, H. E., Mackenzie, E., and Basu, S., "Microwave Equatorial Scintillation Intensity During Solar Maximum," *Radio Science*, Vol. 10, pp. 939-945, Sept.-Oct. 1981.

8.8. Fremouw, E. J., Leadabrand, R. L., Livingston, R. C., Cousins, M. D., Rino, C. L., Fair, B. C., and Long. R. A., "Early Results from the DNA Wideband Satellite Experiment—Complex-Signal Scintillation," *Radio Science*, Vol. 13, pp. 167-187, Jan.-Feb. 1978.

8.9. Rufenach, C. L., "Ionospheric Scintillation by a Random Phase Screen: Spectral Approach," *Radio Science*, Vol. 10, pp. 155-165, Jan.-Feb. 1975.

8.10. CCIR Report 563-2, "Radio-Meteorological Data," in Vol. V, *Propagation in Non-Ionized Media*, Recommendations and Reports of the CCIR—1982, International Telecommunications Union, Geneva, pp. 96-123, 1982.

8.11. Tatarski, V. I., *The Effects of the Turbulent Atmosphere on Wave Propagation*, Nauka, Moscow, 1967.

8.12. CCIR, Report 881, "Effects of Small-Scale Spatial or Temporal Variations of Refraction on Radiowave Propagation," in Vol. V, *Propagation in Non-Ionized Media*, Recommendations and Reports of the CCIR—1982, Geneva, pp. 131-138, 1982.

8.13. CCIR, Report 564-2, "Propagation Data Required for Space Telecommunications Systems," in Vol. V, *Propagation in Non-Ionized Media*, Recommendations and Reports of the CCIR—1982, Geneva, pp. 331-373, 1982.

8.14. Devasirvathm, D. J. M., and Hodge, D. B., "Amplitude Scintillations of Earth-Space Propagation Paths at 2 and 30 GHz," Ohio State Univ., Tech. Report 4299-4, March 1977.

8.15. Titus, J. M., and Arnold, H. W., "Low Elevation Angle Propagation Effects on COMSTAR Satellite Signals," *Bell System Tech. Journal*, Vol. 61, No. 7, pp. 1567-1572, Sept. 1982.

8.16. Ippolito, L. J., "ATS-6 Millimeter Wave Propagation and Communications Experiments at 20 and 30 GHz," *IEEE Trans. on Aerospace and Electronics Systems*, Vol. AES-11, No. 6, pp. 1067-1083, Nov. 1975.

8.17. Cox, D. C., Arnold, H. W, and Hoffman, H. H., "Observation of Cloud-Produced Amplitude Scintillation on 19- and 28-GHz Earth-Space Paths," *Radio Science*, Vol. 16, No. 5, pp. 885-907, Sept.-Oct. 1981.

8.18. Oguchi, T, "Statistical Fluctuation of Amplitude and Phase of Radio Signals Passing through the Rain," *Jour. of the Radio Research Laboratories (Japan)*, Vol. 9, No. 41, Jan. 1962.

8.19. Crane, R. K., "Coherent Pulse Transmission Through Rain," *IEEE Trans. on Antennas and Propagation*, Vol. AP-15, No. 2, March 1967.

8.20. Cox, D. D., Arnold, H. W., and Leck, R. P., "Phase and Amplitude Dispersion for Earth-Satellite Propagation in the 20 to 30 GHz Frequency Range," *IEEE Trans. on Antennas and Propagation*, Vol. AP-28, No. 3, May 1980.

8.21. Theobold, D. M., and Hodge, D. B., "Gain Degradation and Amplitude Scintillation Due to Tropospheric Turbulence," Ohio State Univ., Tech Report 78229-6, Revision 1, May 1978.

8.22. Arnold, H. W., Cox, D. C., and Hoffman, H. H., "Antenna Beamwidth Independence of Measured Rain Attenuation on a 28 GHz Earth-Space Path," *IEEE Trans. on Antennas and Propagation*, Vol. AP-30, No. 2, March 1982.

CHAPTER 9

PROPAGATION EFFECTS ON COMMUNICATIONS SATELLITE LINK PERFORMANCE

Prior chapters have focused on descriptions of the propagation factors that can degrade satellite communications links. Gaseous attenuation, hydrometeor attenuation, depolarization, radio noise, scintillation, and other factors were described, and methods for predicting the effects on communications links were presented.

In this chapter, the performance of satellite communications links, in the presence of one or more propagation effect, will be investigated. Link performance equations for several typical satellite systems, both analog and digital, will be developed, and the parametric variations caused by propagation effects described.

Section 9.1 introduces several parameters which are used in link performance calculations, and Section 9.2 presents detailed link calculations for several types of communications satellite systems.

9.1. COMMUNICATIONS SYSTEM PARAMETERS

Several of the fundamental transmission parameters used to describe radiowave communications were introduced in Chapter 2. Additional parameters and definitions which are necessary for link performance calculations are introduced here, in preparation for the link evaluations which follow. Emphasis is placed on developing an analysis which describes the effects of propagation anomalies on system performance in a most direct manner.

9.1.1. Noise Temperature and Noise Figure

Both equivalent noise temperature, t_e, and noise figure, NF, are used to describe the noise characteristics of a communications system. There are several sources of noise in the communications system. Each amplifier in the receiver system will produce noise power in the information bandwidth, and must be accounted for in a link performance calculation. Other sources include mixers, upconverters/downconverters, switches, combiners, and multiplexers. The system noise produced by these hardware elements is additive to the noise produced in the radiowave transmission path by propagation effects.

The noise figure NF of a noise source of equivalent temperature t_e is given by

$$NF(\text{dB}) = 10 \log_{10} \left(1 + \frac{t_e}{290}\right) \tag{9-1}$$

where t_e is in degrees Kelvin (°K).

Conversely, the equivalent noise temperature of a noise source with a specified noise figure, NF, is found as

$$t_e(°\text{K}) = 290 \left[10^{(NF(\text{dB})/10)} - 1\right] \tag{9-2}$$

The total noise power n in a noise bandwidth of b Hz is

$$n = kt_e b \tag{9-3}$$

where k is Boltzman's constant (−198.6 dBm/°K/Hz).

The noise density n_0 is the noise power present in a 1 Hz bandwidth, i.e.,

$$n_0 = kt_e \tag{9-4}$$

In a communications link system performance calculation, the effective noise temperature of each of the contributing devices is translated to a common reference point, usually the antenna terminals of the receiver, and a total system effective noise temperature is then defined.

Typical system noise temperatures for receive earth stations can range from 20°K to 1500°K (NF = 0.29 to 8 dB), and for satellite receivers from 850°K to 4000°K (NF = 6 to 12 dB).

9.1.2. Figure of Merit

The quality or efficiency of the receiver portions of an earth–satellite communications link is often specified by a figure of merit M, defined as the ratio of antenna gain to the equivalent noise temperature:

$$M(\text{dB}/°\text{K}) = G - 10 \log_{10} t_e \tag{9-5}$$

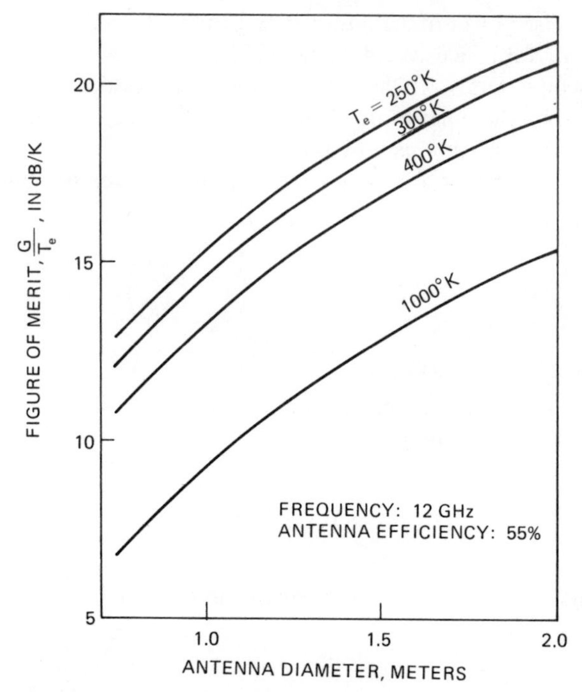

Figure 9-1. Figure of merit for 12 GHz receiver systems as a function of antenna size and system noise temperature.

where G is the antenna gain in dB and t_e is the equivalent noise temperature in °K.

Figure 9-1 shows an example of the figure of merit for a 12 GHz receiver, as a function of antenna diameter and noise temperature. The range of antenna size represents typical values for broadcast satellite service (BSS) applications.

Figure of merit values cover a wide range in operational satellite systems, e.g., from -18.6 dB/°K (INTELSAT IV, C-band) to $+4.4$ dB/°K (TDRS, *K*-band).

9.1.3. Carrier-to-Noise Ratio

The ratio of average wideband carrier power c to the noise power in the same bandwidth, n, is defined as the carrier-to-noise ratio, c/n. The c/n is the primary parameter of interest for defining the overall system performance in a communications system. It can be defined at any point in the link, such as at the uplink receiver antenna terminal, $(c/n)_U$, or the downlink receiver antenna terminal, $(c/n)_D$.

The ratio is sometimes defined as a carrier-to-noise *density* ratio, c/n_0, where $n_0 = kt_e$, the noise density.

For digital communications, the bit energy e_b is more useful than carrier power in describing the performance of the link. The bit energy is obtained from

$$e_b = cT_b \tag{9-6}$$

where c is the carrier power and T_b is the bit duration.

The energy-per-bit to noise density ratio e_b/n_0 is the most frequently used parameter to describe digital communications link performance. e_b/n_0 is related to c/n_0 by

$$\frac{e_b}{n_0} = T_b \frac{c}{n_0} \tag{9-7}$$

This relation allows for a comparison of link performance of both analog and digital modulation techniques, and various transmission rates, for the same link system parameters.

9.1.4. Effective Isotropic Radiated Power

The effective isotropic radiated power, *EIRP*, is a convenient quantitative measure of the transmitting capability in a communications link. The *EIRP* is defined as

$$EIRP = pg \tag{9-8}$$

where p is the transmitted power and g is the antenna gain. For the transmitted power expressed in dBw and the gain in dB,

$$EIRP(\text{dBw}) = P(\text{dBw}) + G(\text{dB}) \tag{9-9}$$

Additional losses in the transmit segment, such as line loss, diplexer loss, or backoff loss, may be included in the EIRP specified for a ground terminal or satellite transmitter system.

The *EIRP* of a 12 GHz, 1 meter diameter antenna (55% efficiency), for example, would be 43.8 dBw for a 1 watt transmitter power, and would be 63.8 dBw for a 100 watt transmitter power, excluding any additional losses.

9.1.5. Percent-of-Time Performance Specifications

It is often necessary, and advantageous, to specify certain communications link system parameters on a statistical basis. This is particularly required when considering parameters affected by propagation factors, since the basic propagation mechanisms are not deterministic, and can only be described on a statistical basis.

Statistically based performance parameters are usually specified on a percent of time basis, that is, the percent of time in a year, or a month, that the parameter is equal to or exceeds a specific vallue. Examples of parameters that are often specified on a percent of time basis are;

Carrier-to-noise ratio: c/n
Video signal-to-noise ratio: s/n
Polarization discrimination: XPD
Protection ratio: C/I

The two most often used time periods for parameter specification are yearly (annual), and worst month. Most propagation effects prediction models and fixed satellite service (FSS) requirements are specified on an annual (8769 hr) basis. Broadcasting services, including the broadcasting satellite service (BSS), usually specify on a worst month (730 hr) basis. The worst month denotes the calendar month where the propagation effect, such as rain attenuation or depolarization, produces the severest degradation on the system parameter. Parameters affected by rain attenuation for example, such as carrier-to-noise ratio or signal-to-noise ratio, would have worst month values in July or August for most regions of the United States, when heavy rain occurrance is most probable.

Figure 9-2 shows a typical method of displaying a link performance parameter specified on a percent of time basis. The parameter is presented on the linear scale of a semi-logarithmic plot, with the percent of time variable placed on the logarithmic scale.

Several terms are used in specifying the percent of time variable, including outage, exceedance, availability, or reliability.

If the percent of time variable is the percent of time the parameter is equaled or exceeded, P, the display represents the *outage* or *exceedance* of the parameter.

If the percent of time variable is $100 - P$, the display represents the *availability* or *reliability* of the parameter.

Table 9-1 shows the annual and monthly outage times, in hours and minutes, corresponding to the range of percentage values of P and $100—P$ typically found in communications link specifications.

For example, a link availability of 99.99%, sometimes referred to as "four nines," corresponds to a link with an expected outage of 0.01%, or 53 minutes, on an annual basis. This value represents the link availability specified in the national telephone system.

The BSS generally specifies link parameters in terms of an outage of 1% of the worst month, corresponding to 7.3 hours or 99% link availability during the worst month.

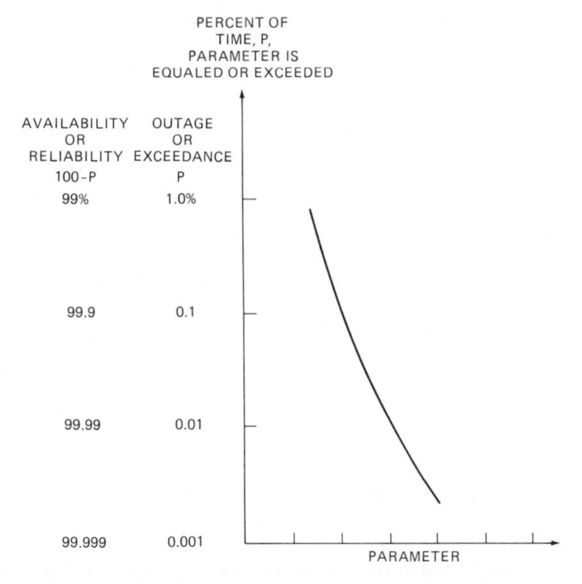

Figure 9-2. Specification of link performance parameters on a percent of time basis.

Most analytical studies and propagation effects measurements are developed on an annual basis. It is often necessary to determine worst month statistics for some specific applications, such as the BSS, from annual statistics, since annual statistics may be the only source of prediction models or measured data available.

The CCIR [9.1] recommends an empirically based relationship of the form

$$P_w \cong 2.94 P_a^{0.87} \tag{9-10}$$

Table 9-1. Annual and Monthly Outage Time for Specified Percent Outage or Availability.

Outage or Exceedance P (%)	Availability or Reliability $100 - P$ (%)	Outage Time: Annual Basis (hr/min per year)	Outage Time: Monthly Basis (hr/min per month)
0%	100%	0 hr	0 hr
10	90	876 hr	73 hr
1	99	87.6 hr	7.3 hr
0.1	99.9	8.76 hr	44 min
0.05	99.95	4.38 hr	22 min
0.01	99.99	53 min	4 min
0.005	99.995	26 min	2 min
0.001	99.999	5 min	0.4 min

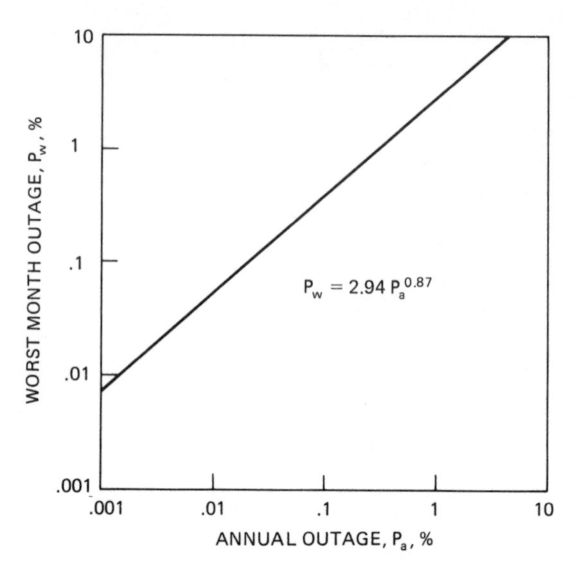

Figure 9-3. CCIR worst month/annual outage statistics relationship.

where P_w is the percent outage for the worst month, and P_a is the percent outage on an annual basis. Conversely,

$$P_a \cong 0.29 P_w^{1.15} \tag{9-11}$$

Figure 9-3 shows a plot of the CCIR relationship for worst month and annual outage percentage values from 0.001% to 10%.

As an example of the application of the CCIR relationship, assume that we wish to determine the margin necessary to maintain an outage no greater than 1% of the worst month, but only annually based rain margin outage measurements are available. The value of the rain margin at the 0.29% point on the annual curve, as seen from the figure, gives the desired 1% worst month value.

This relationship is particularly useful in determining the link margin requirements for worst month applications, such as the BSS, from rain attenuation prediction models such as the global model or the CCIR Model, which are formulated on an annual basis.

9.2. LINK PERFORMANCE IN THE PRESENCE OF PROPAGATION EFFECTS

Propagation effects introduced in the satellite communications link transmission paths (uplink and downlink) are quantitatively determined by including them in the transmission channel portion of the satellite communications system. Prop-

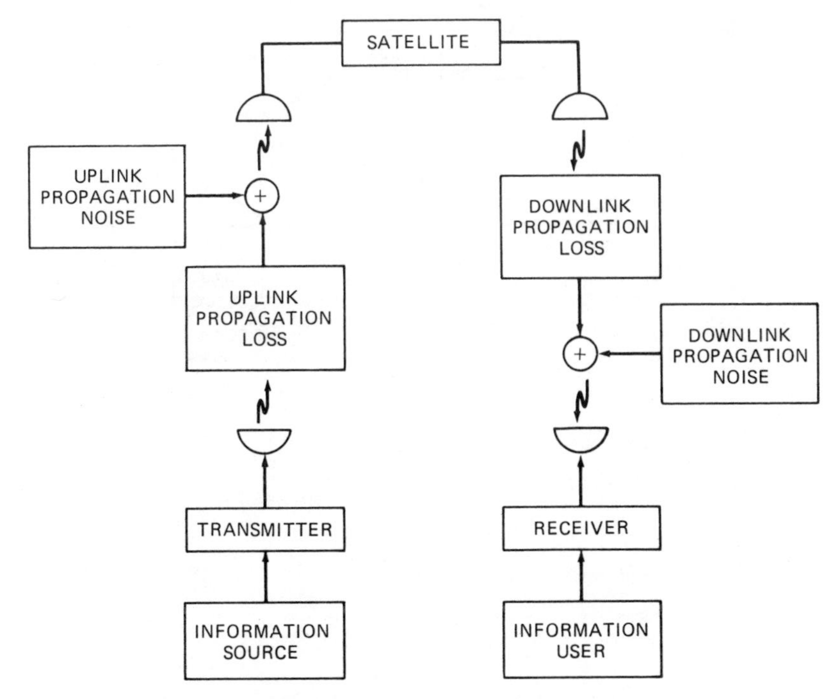

Figure 9-4. Propagation loss and propagation noise in a satellite communications link.

agation losses are introduced in the uplink and the downlink signal paths, and propagation noise is added to the signal at the uplink and downlink, as shown in Figure 9-4.

Propagation loss is the sum of one or more signal power losses caused by effects such as gaseous attenuation, rain or cloud attenuation, scintillation loss, angle of arrival loss, or antenna gain degradation. Propagation noise is the sum of one or more additive noise effects such as noise caused by atmospheric gases, clouds, rain, depolarization, surface emissions, or extraterrestrial sources.

The total system carrier to noise ratio $(c/n)_s$ is determined by developing the system equations for the total link, including the propagation effects parameters. Figure 9-5 defines the parameters used in the link calculations. A subscript preceeded by the letter G is used to denote ground station parameters, and a subscript preceeded by the letter S defines a satellite parameter. Also, parameters given in upper case refer to the parameter expressed in decibels (dB), while lower case refers to the parameter expressed as a number or ratio, in the appropriate units.

Two types of satellite transponders are considered. The first type is the conventional *frequency translation satellite*. The second type is the *on-board processing satellite*, which utilizes on-board detection and remodulation to provide

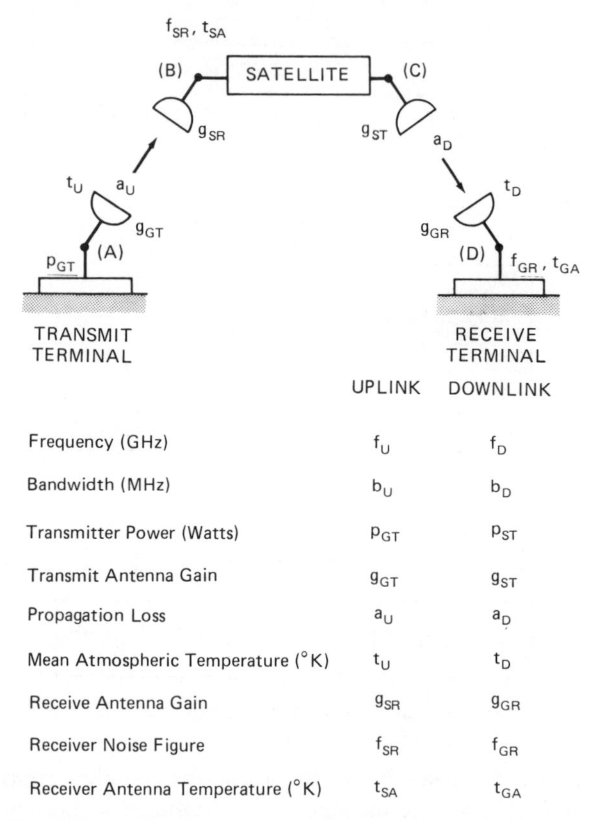

	UPLINK	DOWNLINK
Frequency (GHz)	f_U	f_D
Bandwidth (MHz)	b_U	b_D
Transmitter Power (Watts)	p_{GT}	p_{ST}
Transmit Antenna Gain	g_{GT}	g_{ST}
Propagation Loss	a_U	a_D
Mean Atmospheric Temperature (°K)	t_U	t_D
Receive Antenna Gain	g_{SR}	g_{GR}
Receiver Noise Figure	f_{SR}	f_{GR}
Receiver Antenna Temperature (°K)	t_{SA}	t_{GA}

Figure 9-5. Parameters for link performance calculations.

two essentially independent cascaded (uplink and downlink) communications links.

9.2.1. Frequency Translation Satellite

A conventional frequency translation satellite receives the uplink signal at the uplink carrier frequency f_U, down-converts the information bearing signal to an intermediate frequency f_{IF}, for amplification, up-converts to the downlink frequency, f_D, and, after final amplification, retransmits the signal to the ground. Figured 9-6(a) shows a schematic representation of the conventional frequency translation transponder. An alternate version, the direct frequency translation transponder, is shown in Figure 9-6(b). In the direct transponder, the uplink frequency is converted directly to the downlink frequency, and after one or more stages of amplification, retransmitted to the ground.

No processing is done on board the frequency translation satellite. Signal

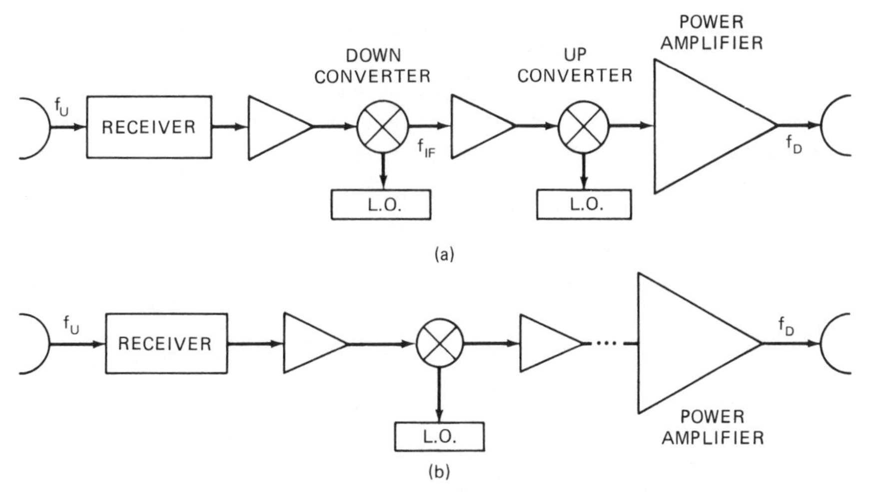

Figure 9-6. Frequency translation transponder.

degradations and noise introduced on the uplink are translated to the downlink, and the total performance of the system is dependent on both links.

The ground transmit terminal *EIRP*, using the nomenclature of Figure 9-5, is

$$EIRP_G = p_{GT} g_{GT} \tag{9-12}$$

The carrier power received at the satellite antenna terminals, point B in Figure 9-5, is

$$c_{SR} = p_{GT} g_{GT} l_U a_U g_{SR} \tag{9-13}$$

where l_U is the uplink free space path loss [see Equation (2-9)], a_U is the uplink propagation loss, and g_{GT} and g_{SR} are the transmit and receive antenna gains, respectively.

The noise power at the satellite antenna, point B, is the sum of three components, i.e.,

Noise Power at Satellite Antenna	=	Uplink Propagation Noise	+	Satellite Receiver Antenna Noise	+	Satellite Receiver System Noise

The three components are

$$n_{SR} = kt_U[1 - a_U]b_U + kt_{SA}b_U + k290[f_{SR} - 1]b_U \tag{9-14}$$

where k is Boltzman's constant, b_U is the uplink information bandwidth, t_{SA} is the satellite receiver antenna temperature, f_{SR} is the satellite receiver noise figure [see Equation (9-2)], and t_U is the mean temperature of the uplink atmospheric path [see Equation (7-5)].

Therefore

$$n_{SR} = kb_U[t_U(1 - a_U) + t_{SA} + 290(f_{SR} - 1)] \tag{9-15}$$

The *uplink* carrier-to-noise ratio at point B is then given by

$$\left(\frac{c}{n}\right)_U = \frac{c_{SR}}{n_{SR}} = \frac{p_{GT}g_{GT}l_U a_U g_{SR}}{kb_U[t_U(1 - a_U) + t_{SA} + 290(f_{SR} - 1)]} \tag{9-16}$$

The *downlink* carrier-to-noise ratio is found by following the same procedure, and using the equivalent downlink parameters as defined in Figure 9-5. Thus, at point D,

$$c_{GR} = p_{ST}g_{ST}l_D a_D g_{GR} \tag{9-17}$$

$$n_{GR} = kb_D[t_D(1 - a_D) + t_{GA} + 290(f_{GR} - 1)] \tag{9-18}$$

and

$$\left(\frac{c}{n}\right)_D = \frac{c_{GR}}{n_{GR}} = \frac{p_{ST}g_{ST}l_D a_D g_{GR}}{kb_D[t_D(1 - a_D) + t_{GA} + 290(f_{GR} - 1)]} \tag{9-19}$$

The downlink transmit power p_{ST} for a frequency translation satellite, will contain both the desired carrier component, c_{ST} and noise introduced by the uplink and by the satellite system itself, n_{ST}. That is

$$p_{ST} = c_{ST} + n_{ST} \tag{9-20}$$

Also, since there is no on-board processing or enhancement of the information signal,

$$\frac{c_{ST}}{n_{ST}} = \frac{c_{SR}}{n_{SR}} = \left(\frac{c}{n}\right)_U \tag{9-21}$$

that is, the satellite output carrier-to-noise ratio is equal to the satellite input carrier-to-noise ratio. (This assumes that all noise introduced by the satellite system is accounted for by f_{SR}.)

Combining Equations (9-20) and (9-12),

$$c_{ST} = \frac{p_{ST}}{1 + [1/(c/n)_U]} \tag{9-22}$$

and

$$n_{ST} = \frac{c_{ST}}{(c/n)_U} = \frac{p_{ST}}{1 + (c/n)_U} \tag{9-23}$$

The carrier power received at the ground receiver, point D, c'_{GR}, will be, from Equation (9-22),

$$c'_{GR} = \frac{p_{ST}}{1 + [1/(c/n)_U]} g_{ST} l_D a_D g_{GR} \tag{9-24}$$

or, from Equation (9-17),

$$c'_{GR} = \frac{c_{GR}}{1 + [1/(c/n)_U]} \tag{9-25}$$

The total noise power received on the ground, n'_{GR}, will be the sum of the noise introduced on the downlink, Equation (9-18), and the noise transferred from the uplink, Equation (9-23), that is,

$$n'_{GR} = n_{GR} + n_{ST} g_{ST} l_D a_D g_{GR} \tag{9-26}$$

Using Equation (9-23),

$$n'_{GR} = n_{GR} + \frac{p_{ST}}{1 + (c/n)_U} g_{ST} l_D a_D g_{GR} \tag{9-27}$$

Comparing with Equation (9-17),

$$n'_{GR} = n_{GR} + \frac{c_{GR}}{1 + (c/n)_U} \tag{9-28}$$

Equations (9-25) and (9-28) show the effect of uplink degradations on the downlink signal and noise, respectively.

Finally, the *total system* carrier-to-noise ratio $(c/n)_S$ is found as

$$\left(\frac{c}{n}\right)_S = \frac{c'_{GR}}{n'_{GR}} = \frac{c_{GR}/\{1 + [1/(c/n)_U]\}}{n_{GR} + (c_{GR})/[1 + (c/n)_U]} \tag{9-29}$$

Rearranging terms and expressing the results in terms of the uplink and downlink carrier-to-noise ratios,

$$\left(\frac{c}{n}\right)_S = \frac{(c/n)_U\,(c/n)_D}{1 + (c/n)_U + (c/n)_D} \tag{9-30}$$

where $(c/n)_U$ and $(c/n)_D$ are given by Equations (9-16) and (9-19), respectively.

Note that for both $(c/n)_U$ and $(c/n)_D >> 1$, Equation (9-30) reduces to,

$$\left(\frac{c}{n}\right)_S \cong \frac{1}{\dfrac{1}{(c/n)_U} + \dfrac{1}{(c/n)_D}} \tag{9-31a}$$

or

$$\frac{1}{\left(\dfrac{c}{n}\right)_S} \cong \frac{1}{(c/n)_U} + \frac{1}{(c/n)_D} \tag{9-31b}$$

which is the result most often found in textbooks for the total system carrier-to-noise ratio of a satellite link.

The complete result, given by Equation (9-30), is more accurate for general link analysis, particularly when small variations in system parameters or propagation effects are being evaluated.

The total system carrier-to-noise *density* $(c/n_0)_S$, can easily be shown to be of the same form as Equation (9-30), i.e.,

$$\left(\frac{c}{n_0}\right)_S = \frac{(c/n_0)_U\,(c/n_0)_D}{1 + (c/n_0)_U + (c/n_0)_D} \tag{9-32}$$

with

$$\left(\frac{c}{n_0}\right)_U = \frac{p_{GT}\,g_{GT}\,l_U\,a_U\,g_{SR}}{k[t_m(1 - a_U) + t_{SA} + 290(f_{SR} - 1)]} \tag{9-33}$$

and

$$\left(\frac{c}{n_0}\right)_D = \frac{p_{ST}g_{ST}l_D a_D g_{GR}}{k[t_m(1 - a_D) + t_{GA} + 290(f_{GR} - 1)]} \tag{9-34}$$

For digital communications links, the energy-per-bit to noise density ratio of the total system, $(e_b/n_0)_T$, can be found by inserting

$$\left(\frac{e_b}{n_0}\right) = T_b\left(\frac{c}{n_0}\right) \tag{9-35}$$

into Equation (9-32), where T_b is the bit duration [see Equation (9-7)].

For the conditions given for Equation (9-31), i.e., both $(c/n_0)_U$ and $(c/n_0)_U$ >> 1, the total system (e_b/n_0) reduces to

$$\left(\frac{e_b}{n_0}\right)_S = \frac{1}{\dfrac{1}{(e_b/n_0)_U} + \dfrac{1}{(e_b/n_0)_D}} \tag{9-36a}$$

or

$$\frac{1}{(e_b/n_0)_S} = \frac{1}{(e_b/n_0)_U} + \frac{1}{(e_b/n_0)_D} \tag{9-36b}$$

The probability of error for the total digital link is determined from the total energy-per-bit to noise density described above.

It should be reemphasized that the parameters and ratios presented in this section are expressed as numerical values, not as dB values.

9.2.2. On-Board Processing Satellite

A satellite which contains on-board processing provides two essentially independent cascaded communications links for the uplink and downlink. Figure 9-7 shows a schematic block diagram of a generalized on-board processing satellite transponder. The information signal on the uplink carrier, f_U, is demodulated, and the baseband signal is amplified and enhanced by one or more signal processing techniques such as reformatting or error-correcting coding. The enhanced baseband signal may then be introduced on the downlink with a completely different modulation scheme. Degradations added to the uplink may be nearly fully compensated for by the on-board processing, and are not transferred to the downlink.

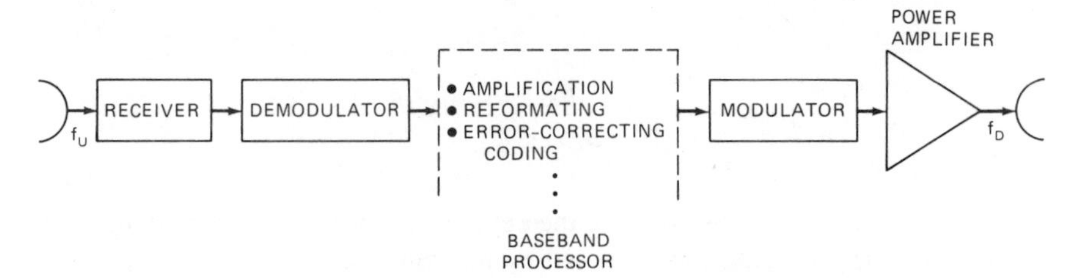

Figure 9-7. On-board processing satellite transponder.

The downlink carrier-to-noise ratio is essentially independent of the uplink carrier-to-noise ratio over the operating range of the transponder. The link equations developed in the previous section for $(c/n)_U$ and $(c/n)_D$, Equations (9-16) and (9-19), respectively, are applicable for the on-board processing satellite system. The remainder of the analysis, however, i.e., Equations (9-20) through the end of the section, do not apply to the on-board processing satellite system.

Each link must be handled separately, and the resulting system performance of the total link is evaluated for each link individually, depending on the particular configuration of the satellite system.

9.2.3. Effects of Path Attenuation on System Performance

The results given in the previous sections give the total system performance as a function of all of the link variables. It is not obvious from the $(c/n)_S$ results, i.e., Equations (9-30) and (9-31), how propagation degradations will quantitatively affect overall link performance. Path attenuations a_U and a_D, and propagation noise temperatures t_U and t_D, are found in both the numerator and the denominator of the $(c/n)_S$ equations, and their relative contributions will depend to a large extent on the values of the other parameters of the system such as transmit powers, antenna gains, noise figures, information bandwidths, etc.

Each particular satellite system must be analyzed on an individual basis, and propagation degradations evaluated by a sensitivity analysis for specified values of the other system parameters. Relative contributions of uplink and downlink propagation effects, and the overall performance of the system in the presence of propagation degradations can then be determined. The system design could then be optimized to achieve a desired level of availability or performance by adjusting those system parameters that can be changed.

A specific case study will be presented here to demonstrate several important characteristics of propagation effects on system performance.

The Communications Technology Satellite, CTS, launched in January 1976 [9.2], provides an excellent example of the wide range of operational configu-

Table 9-2. Link Parameters for the Communications Technology Satellite (CTS).

	Uplink	Downlink
Frequency (GHz)	14.1	12.1
Bandwidth (MHz)	30	30
Transmit Power (watts)	100–1000	20/200
Transmit antenna gain (dB)	54	36.9
Receive antenna gain (dB)	37.9	52.6
Receiver noise figure (dB)	8	3
Receive antenna temp. (°K)	290	50
Free space path loss (30° elevation angle) (dB)	207.2	205.8

rations available in a conventional frequency translation satellite. The CTS was an experimental communications satellite, operating in the 14/12 GHz frequency bands. It contained a high power (200 watt) downlink transmitter for direct broadcast experiments, and a lower power (20 watt) downlink transmitter found in typical point-to-point fixed satellite applications.

Table 9-2 lists the CTS system parameters for the uplink and downlink. A 30° elevation angle for both the uplink and the downlink was assumed, resulting in the free space pass loss values listed in the table.

The CTS operated at a downlink transmitt power of either 20 watts or 200 watts, and uplink powers from 100 watts to 1000 watts, depending on specific ground terminal capabilities. A large range of uplink/downlink transmit power combinations were possible with the CTS, and the effects of propagation losses (primarily rain) differed with each combination.

Four CTS modes of operation will be considered here. Table 9-3 lists the uplink and downlink transmitt powers for each mode, along with the "clear sky" values of $(C/N)_U$, $(C/N)_D$, and $(C/N)_S$, as calculated from Equations (9-16), (9-17), and (9-30), respectively. The clear sky values correspond to the condition of no propagation degradations, i.e., A_U and A_D are both 0 dB.

Table 9-3. Clear Sky Carrier-to-Noise Ratios for the Communications Technology Satellite (CTS).

	Mode 1	Mode 2	Mode 3	Mode 4
Uplink transmit power (watts)	100	100	1000	1000
Downlink transmit power (watts)	200	20	20	200
Uplink, $(C/N)_U$ (dB)	25.9	25.9	35.9	35.9
Downlink, $(C/N)_D$ (dB)	35.2	25.2	25.2	35.2
System, $(C/N)_S$ (dB)	25.5	22.5	24.9	32.6

Note that Mode 1 is an *uplink limited* system, where the uplink is the weaker link by about 10 dB. Mode 3 is a *downlink limited* system by about the same margin. Modes 2 and 4 are about even between the uplink and the downlink, hence $(C/N)_S$ is about 3 dB less than either link. Mode 4 is the best link in terms of expected overall system performance with a value of 32.6 dB for $(C/N)_S$. Mode 3 is representative of a typical fixed satellite point-to-point link, which is downlink limited and operates with a rain margin of about 15 dB above the receiver threshold value of 10 dB. Mode 4 is typical for the satellite portion of a high power direct broadcast (DBS) link. The ground receiver antenna gain on the downlink would be lower by about 13 dB for a typical direct home receiver system, however.

Figures 9-8 through 9-11 show the effects of path attenuation on system performance for the four CTS modes, respectively. The figures show the reduction in total system carrier-to-noise ratio, $\Delta(C/N)_S$, as a function of downlink path attenuation A_D and uplink path attenuation A_U. The plotted curves are for the range of A_D from 0 to 30 dB, with fixed values of A_U in 5 dB increments as indicated on the plots. The dashed horizontal line on each figure indicates the value of $\Delta(C/N)_S$ at the receiver threshold value of 10 dB carrier-to-noise ratio.

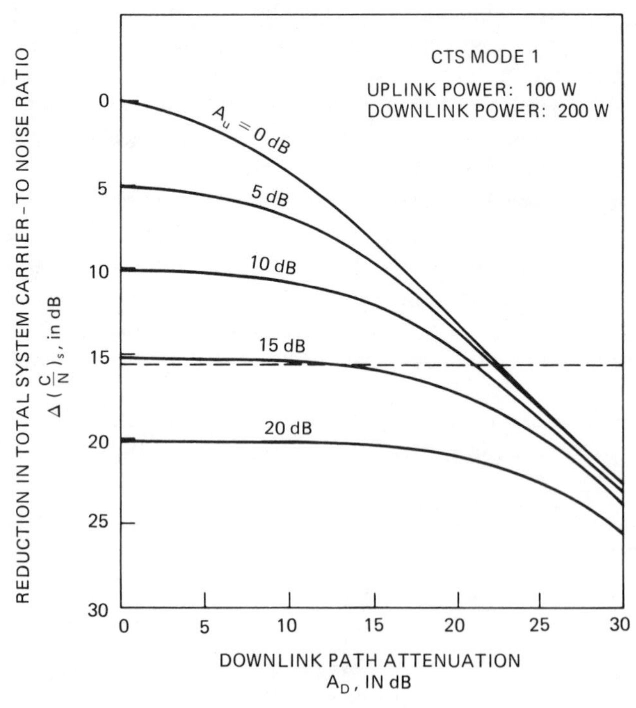

Figure 9-8. Effects of path attenuation on system performance—CTS Mode 1.

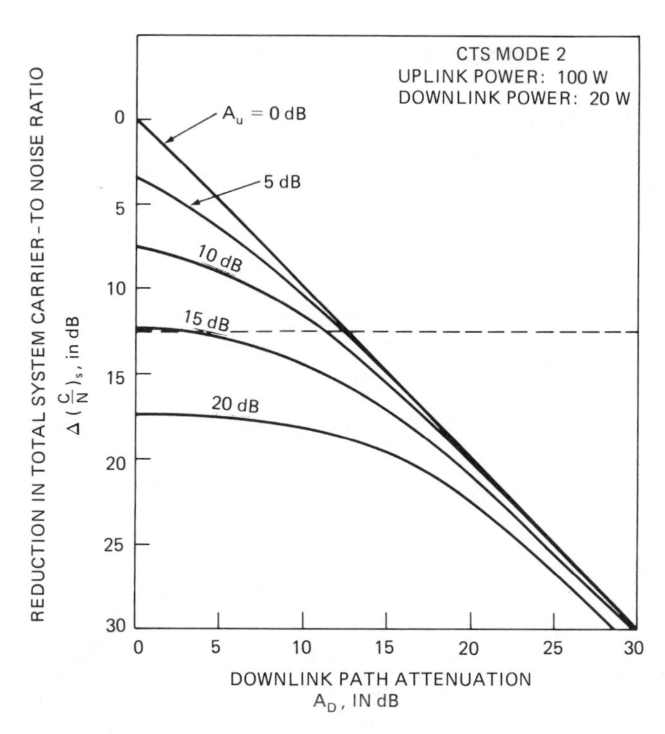

Figure 9-9. Effects of path attenuation on system performance—CTS Mode 2.

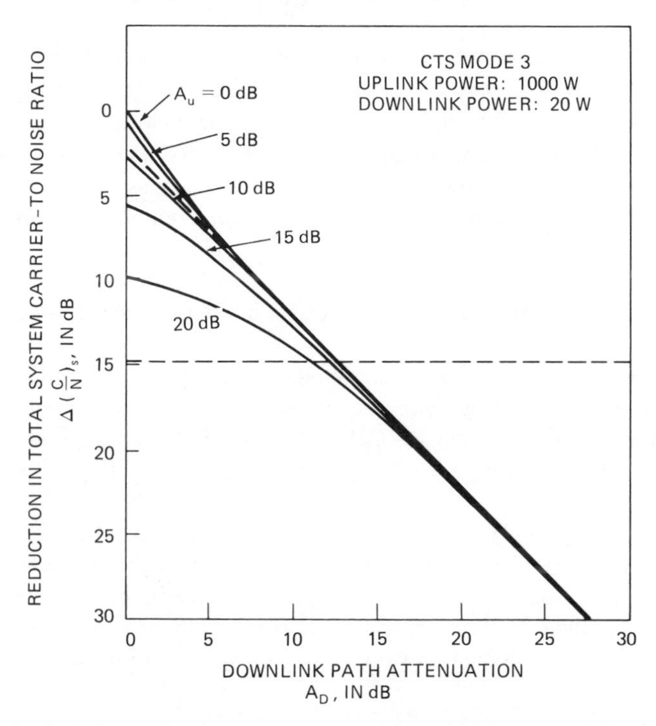

Figure 9-10. Effects of path attenuation on system performance—CTS Mode 3.

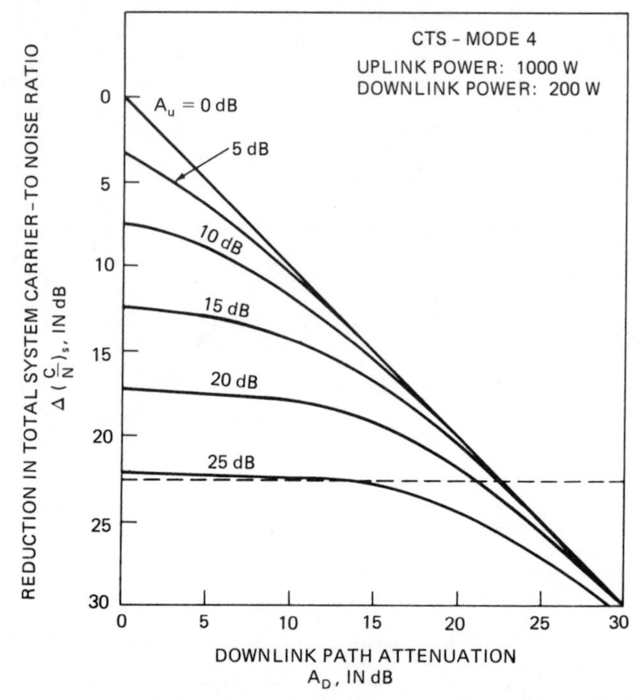

Figure 9-11. Effects of path attenuation on system performance—CTS Mode 4.

The figures highlight the differences in the characteristics of the performance for the various modes. For Mode 1 (Figure 9-8), the curves show a gradual dropoff of $(C/N)_S$ as the downlink path attenuation increases, for low values of uplink attenuation. With no uplink attenuation, for example, a 10 dB downlink attenuation results in only a 4.3 dB reduction in $(C/N)_S$, and a 20 dB downlink attenuation results in a 13 dB reduction in $(C/N)_S$. Conversely, an uplink attenuation, with no downlink attenuation, results in a nearly one-for-one reduction in $(C/N)_S$.

Mode 2 (Figure 9-9) shows a much sharper dropoff of $(C/N)_S$ as the downlink attenuation increases. The 10 dB downlink attenuation example given above results in a 9.9 dB reduction in $(C/N)_S$, the 20 dB level results in a 19.8 dB reduction in $(C/N)_S$; i.e., essentially a one-for-one dropoff is observed with downlink attenuation.

Mode 4 (Figure 9-11), is very similar in performance to Mode 2, as seen by the plots. The major difference is in the location of the receiver threshold. Mode 4 is about 10 dB "better," since it operates with higher transmit powers in both the uplink and the downlink. Note that in Mode 4, the link will operate above threshold even with attenuation values of 25 dB on the uplink and 10 dB on the downlink. Thus a total of 35 dB of attenuation is present on the links, but the $(C/N)_S$ is degraded by less than the 22.6 dB needed to maintain operation!

Mode 3 performance (Figure 9-10) is perhaps the most unusual of all. For low levels of uplink attenuation (A_U = 0–5 dB), the reduction in $(C/N)_S$ is actually *greater* than the downlink attenuation. A 5 dB downlink attenuation produces a 6.6 dB reduction in $(C/N)_S$; a 10 dB downlink attenuation produces a 12 dB reduction in $(C/N)_S$. The system performance of this downlink limited mode is very sensitive to downlink attenuation, and the increased values occur because total system degradation is the sum of the signal loss (attenuation) and the increase in noise power due to the attenuating path. The noise power is 1.6 dB and 2 dB, respectively, for the examples given above.

The results above vividly point out the need to evaluate each specific communications link configuration individually, since small changes in system parameters can substantially change system performance in the presence of path attenuation, and may produce results which are not anticipated in the original system design.

REFERENCES

9.1. CCIR, Report 723-1, "Worst-Month Statistics," Recommendations and Reports of the CCIR, 1982, Volume V, Geneva, 1982.

9.2. Evans, W. M., Davies, N. G., and Hawersaat, W. H., "The Communications Technology Satellite (CTS) Program," in *Communications Satellite Systems: An Overview of the Technology*, R. G. Gould and Y. F. Lum, Eds., IEEE Press, New York, 1976, pp. 13–18.

CHAPTER 10
RESTORATION TECHNIQUES FOR OVERCOMING SEVERE ATTENUATION

A space communications system which is subject to weather dependent path attenuation can be designed to operate at an acceptable performance level by allowing adequate power margins on the uplink and the downlink segments. This can be accomplished directly by increasing antenna size, or increasing the RF transmit power, or both. Typically, power margins of 5–10 dB at *C*-band and 10–15 dB at *K*-band can be relatively easily achieved with reasonably sized antennas and with RF power within allowable levels. RF power levels are most likely constrained by prime power limitations on the satellite, and by radiated power limitations on the ground fixed by international agreement.

If the expected path attenuations exceed the power margins available, and this can easily occur at K_U and K_A band for many regions of the earth, then additional methods must be considered to overcome the severe attenuation conditions and restore acceptable performance on the links.

In this chapter we will review several of the more promising restoration techniques available to the systems designer for overcoming severe attenuation conditions on earth–space links. These techniques include site diversity, power control, orbital diversity, spot beams, frequency diversity, and adaptive forward error correction. Many of these techniques have been employed on space communications systems, and some have been found to give improved performance (sometimes very spectacular!) for links subject to severe rain attenuation and other propagation degradations.

10.1. SITE DIVERSITY

Site diversity is the general term used to describe the utilization of two (or more) geographically separate ground terminals in a space communications link to overcome the effect of path attenuation during intense rain periods. Site diversity, also referred to as path diversity or space diversity, can improve overall satellite link performance by taking advantage of the limited size and extent of

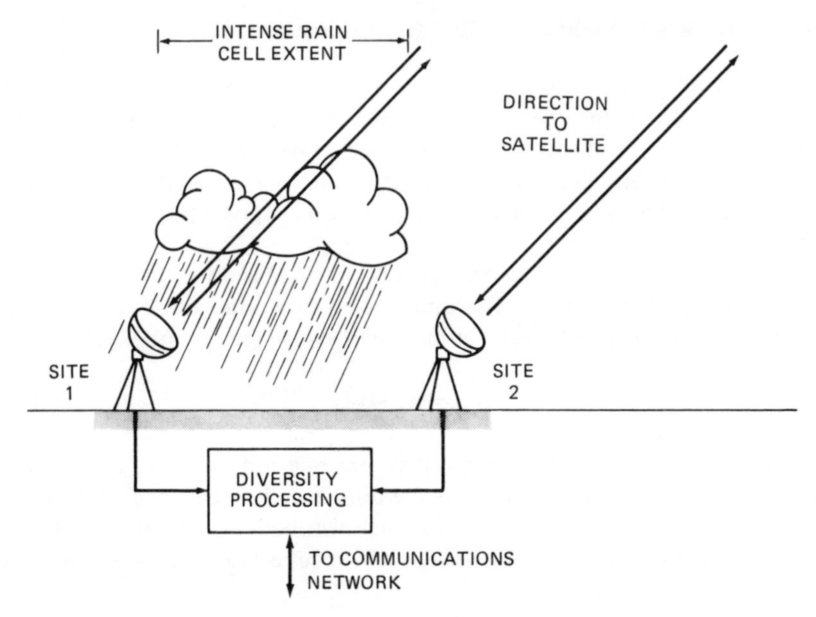

Figure 10-1. Site diversity concept.

intense rain cells. With sufficient physical separation between ground terminals, the probability of a given rain attenuation level being exceeded at both sites is much less than the probability of that attenuation level being exceeded at a single site.

Figure 10-1 shows the concept of site diversity for the two ground terminal case. Heavy rain usually occurs within cellular structures of limited horizontal and vertical extent. These rain cells could be just a few kilometers in horizontal and vertical size, and tend to be smaller as the intensity of the rain increases. If two ground stations are separated by at least the average horizontal extent of the rain cell, then the cell is unlikely to intersect the satellite path of both ground terminals at any given time. As shown in the figure, the rain cell is in the path of Site 1, while the path of Site 2 is free of the intense rain. As the rain cell moves through the region, it may move into the Site 2 path, but the Site 1 path would then be clear.

The downlink received signals from the two terminals are brought to a single location (which could be at one of the terminals), where the signals are compared and a decision process is implemented to select the ''best'' signal for use in the communications system. Transmitted (uplink) information likewise can be switched between the two terminals using a decision alogrithm based on the downlink signal, or on other considerations. (Diversity processing techniques are discussed further in Section 10.1.4).

10.1.1. Diversity Gain and Diversity Improvement

The impact of site diversity on system performance can be quantitatively defined by considering the attenuation statistics associated with a single terminal and with diversity terminals for the same rain conditions. *Diversity gain* is defined as the difference between the path attenuations associated with the single terminal and a diversity of modes of operation for a given percentage of time [10.1].

Figure 10-2 shows how diversity gain G_D is graphically represented on the attenuation distribution plots. The difference between the single terminal and joint terminal attenuation values, at the same percentage of time, i.e.,

$$G_D = A_S - A_J \tag{10-1}$$

is the diversity gain for the system at that percentage of time. For the example shown on the figure, G_D at 0.01% is the difference between 9 dB (single terminal distribution) and 3 dB (joint terminal distribution), or 6 dB. That is, a single terminal operating on this link would require a 9 dB power margin to

DIVERSITY GAIN

"THE DIFFERENCE BETWEEN THE PATH ATTENUATIONS ASSOCIATED WITH THE SINGLE TERMINAL AND DIVERSITY MODES OF OPERATION FOR A GIVEN PERCENTAGE OF TIME."

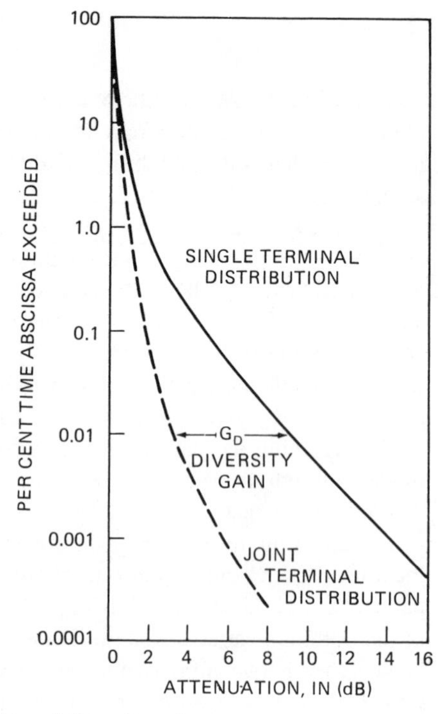

Figure 10-2. Definition of diversity gain.

maintain a 0.01% outage for the link, while two terminals operating in a diversity configuration could maintain that same reliability with only 3 dB power margins. The additional 6 dB margin is a gain in the system power budget, hence the use of the term diversity gain.

The 6 dB power savings could be realized by reducing the antenna size at the two terminals by one-half, or by reducing the transmit power by a factor of 4. The cost savings resulting from these reductions would, of course have to be compared with the additional costs of adding the second terminal to determine the overall economic benefits of adding diversity operation to the system.

The primary parameter in a diversity configuration which determines the amount of improvement obtained by diversity operation is the site separation. Figure 10-3 shows an idealized presentation of the dependence of diversity gain G_D on the site separation distance d for the two terminal diversity case. Each curve corresponds to a fixed value of single terminal attenuation, with $A_2 > A_1$, $A_3 > A_2$, etc. As site separation distance is increased, diversity gain will also increase, up to about the average horizontal extent of the intense rain cell. At separation distances well beyond the average horizontal extent, there is little improvement in diversity operation. If the site separation distance is too great, diversity gain can actually decrease, since a second cell could become involved in the propagation paths.

Figure 10-4 shows the variation of diversity gain with single terminal attenuation, for fixed values of site separation d, where $d_2 > d_1$, $d_3 > d_2$, etc. Beyond the "knee" of the curves, the diversity gain increases nearly one-for-one with attenuation, for a given value of d. As d is increased, the diversity gain increases and approaches the ideal (but unrealizable) condition where the

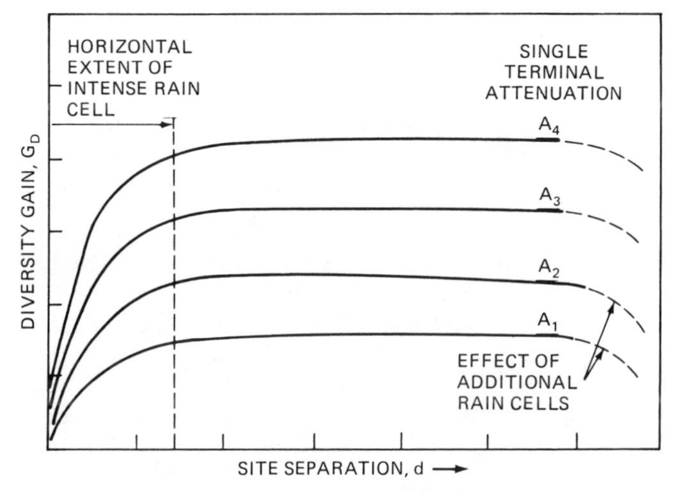

Figure 10-3. Dependence of diversity gain on site separation (idealized).

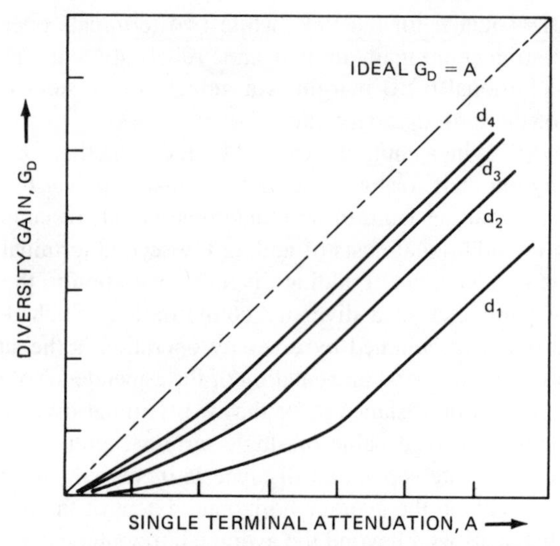

Figure 10-4. Diversity gain and single terminal attenuation (idealized).

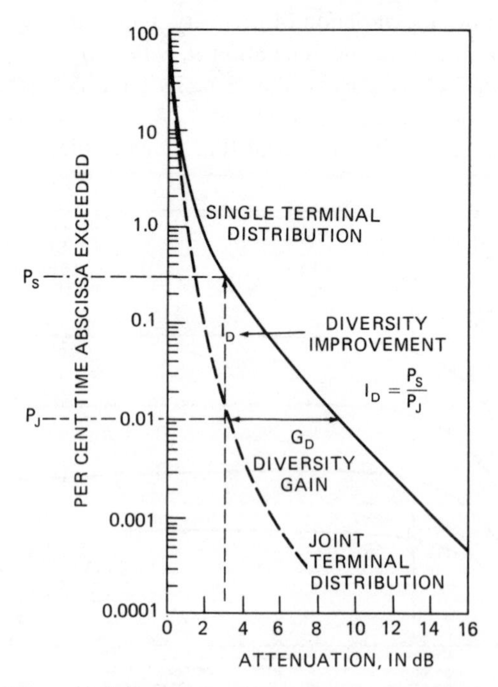

Figure 10-5. Definition of diversity improvement.

attenuation is completely compensated for by the diversity effect, i.e., $G_D = A$. If d is increased too far, the diversity gain can, as described above, begin to decrease (not shown on this figure).

Diversity performance can also be described in terms of outage times by a factor known as *diversity improvement*, described by Figure 10-5. Diversity improvement is defined as

$$I_D = \frac{P_S}{P_J} \tag{10-2}$$

where P_S is the percent of time associated with the single terminal distribution, and P_J is the percent of time associated with the joint terminal distribution, both at the same value of attenuation. For the example shown on the figure, $P_S = 0.3$ and $P_J = 0.01$, both at the attenuation value of 3 dB. The diversity improvement is therefore, $I_D = 30$. That is, a single terminal operating in this link with a 3 dB power margin would suffer an outage of 1577 minutes per year (0.3%). Two terminals operating in a diversity configuration, both with the same 3 dB power margins, would have a joint outage time of only 53 minutes per year (0.01%), an improvement of 30 to 1! Diversity improvement factors of over 100 are not unusual, particularly in areas of intense thunderstorm occurrence.

10.1.2. Diversity Measurements

Extensive direct measurements of site diversity characteristics on operating links are available from radiometer, satellite beacon, and radar based experiments. Measurements are typically accomplished by observing direct path attenuation at two or more sites and then developing joint attenuation statistics and distributions.

Figure 10-6 shows an example of dual site diversity measurements at INTELSAT facilities in West Virginia, using 11.6 GHz radiometers [10.2]. The measurements were taken over a one year period. The site separation was 35 km and the elevation angle was 18°. Single site distributions for the two sights, Etam and Lenox, are shown on the figure, as well as the joint (diversity) distribution. Note that the two single site distributions differ slightly; however, there is a significant improvement in diversity performance throughout the range of the measurements.

Figure 10-7 shows the diversity gain for the measurement period, referenced to both the Etam and the Lenox single site distributions.* The results show that, for a link availability of 99.95%, power margins of 8.8 dB and 7.3 dB at Etam

*The dashed curve labeled ''Hodge Prediction'' is discussed in the next section.

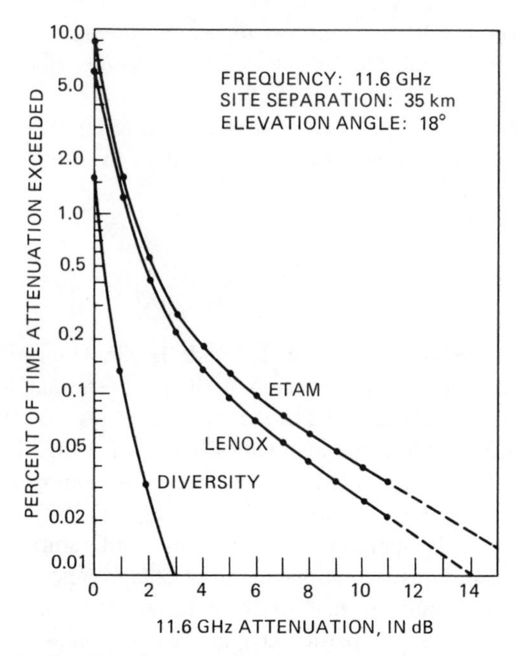

Figure 10-6. Site diversity measurements at 11.6 GHz in West Virginia.

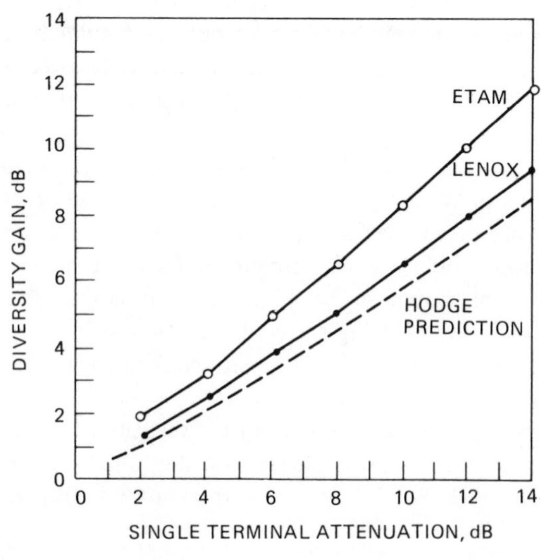

Figure 10-7. Diversity gain measurements for the West Virginia experiment.

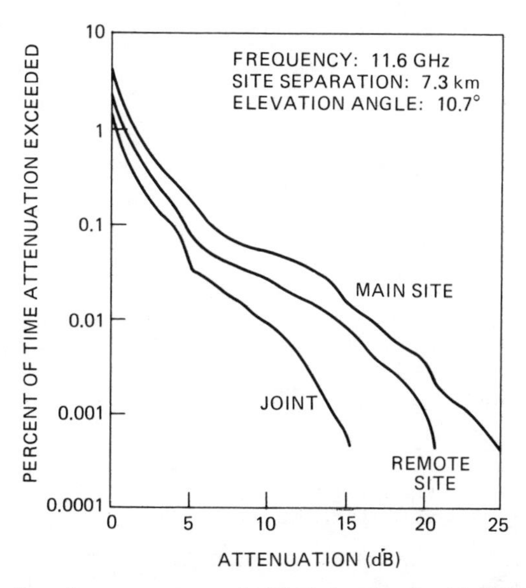

Figure 10-8. Site diversity measurements in Virginia using the 11.6 GHz SIRIO beacon.

and Lenox, respectively, would have been required for single site operation. With diversity operation, the power margin is reduced to 1.6 dB, resulting in diversity gain values of 7.2 dB and 5.7 dB for Etam and Lenox, respectively.

Figure 10-8 shows the attenuation distributions for dual site diversity measurements made at Blacksburg, Virginia, using the 11.6 GHz beacon on the SIRIO satellite [10.3]. The measurements were for a one year period, with a site separation of 7.3 km, at an elevation angle of 10.7 degrees. The measurements show that, at a link availability of 99.95%, power margins of 11.8 dB and 7 dB at the main and remote sites, respectively, would have been required for single site operation. With diversity operation, the power margin reduces to 4.5 dB, corresponding to diversity gain values of 7.3 dB and 2.5 dB for the two sites, respectively. Note also that a 99.99% availability can be achieved with a 10 dB power margin in the diversity mode, while the margins required for single site operation are 17 dB and 14.5 dB for the main and remote sites, respectively.

Dramatic diversity improvement can be achieved by the use of *three* ground stations, particularly in intense thunderstorm areas, where large rain attenuations can occur over significant percentages of the year. This was vividly demonstrated in an experiment carried out in Tampa, Florida using beacons of the COMSTAR satellites [10.4]. Three sites, the University of Florida (labeled U), Lutz (L), and Sweetwater (S), arranged as shown on the plot of Figure 10-9(a), constituted the Tampa triad experiment.

Site operations were 11.3, 16.1, and 20.3 km. The 19.06 GHz beacons of the COMSTAR D2 and D3 satellites were monitored at all three sites. Since

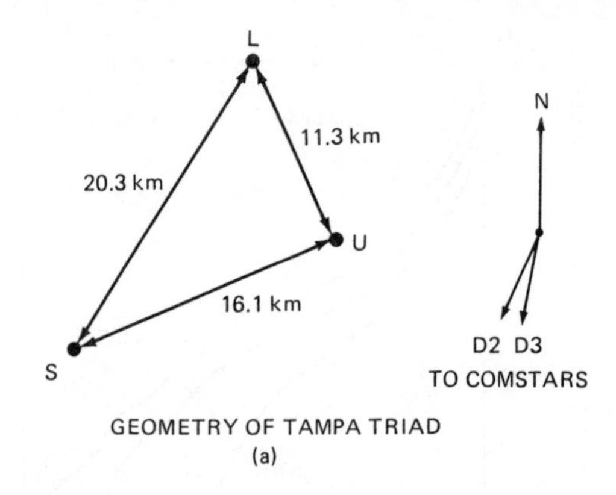

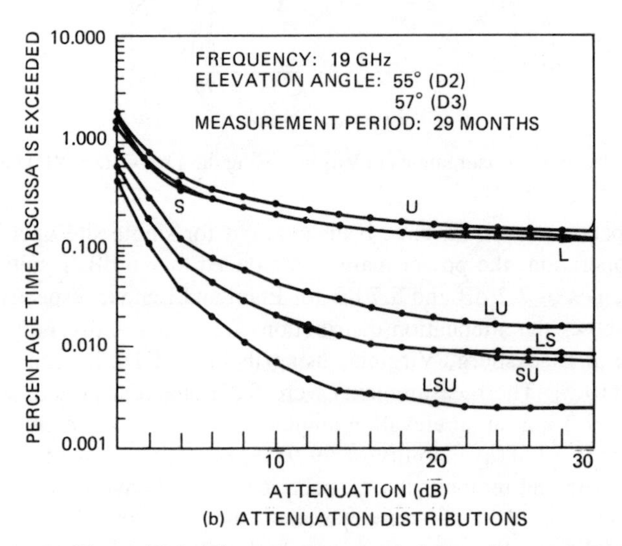

Figure 10-9. Three site Tampa triad diversity measurements.

COMSTAR D2 was located at 95° West longitude, and COMSTAR D3 at 87° West longitude, the elevation angles to the satellites differed slightly; 55° for D2 and 57° for D3. The azimuth angles to the satellites differed by about 16 degrees, as shown on the figure.

Single site and joint attenuation distributions for a 29 month measurement period are shown in Figure 10-9(b). The single site distributions, labeled S, U, and L, demonstrate the severe nature of rain attenuation in the Tampa area. The distributions are nearly flat above about 5 dB of attenuation, indicating that most of the attenuation is due to heavy, thunderstorm associated rain. Attenuation exceeded 30 dB for 0.1% (526 minutes annually), and reliable communications could not be achieved for more than about 99.7% of an average year (26 hours annual outage) with a 10 dB power margin. Link availabilities better than 99.9% could not be achieved at any power margin, with single site operation.

Two site diversity operation provides some improvement, as seen by the LU, LS, and SU distributions. Link availabilities of 99.96% (LU) and 99.98% (LS or SU) can be achieved with a 10 dB power margin.

Three site diversity provides impressive improvement, as seen by the LSU distribution. A 99.99% availability can be achieved with about a 9 dB power margin. The diversity improvement factor I for the Tampa triad at the 10 dB attenuation level was 43.

The Tampa triad experiment highlighted the utility of site diversity, particularly three site diversity, for the restoration of system performance in a severe rain environment.

10.1.3. Diversity System Performance and Design

The improvements possible with diversity operation on earth–space links are dependent on a number of factors. The site separation distance is perhaps the most critical. Diversity gain increases as site separation increases (see Figure 10-3) up to a distance of about 10 km, and beyond that value there is very little increase as diversity gain increases.

Baseline orientation with respect to the propagation path is also an important consideration in configuring a diversity system. If the angle between the baseline and the surface projection of the path to the satellite is 90°, the probability of both paths passing through the same rain cell is greatly reduced. Figure 10-10(a) shows the case of optimum site location, where the baseline orientation angle Δ is 90°. Figure 10-10(b) shows the least desirable configuration, where

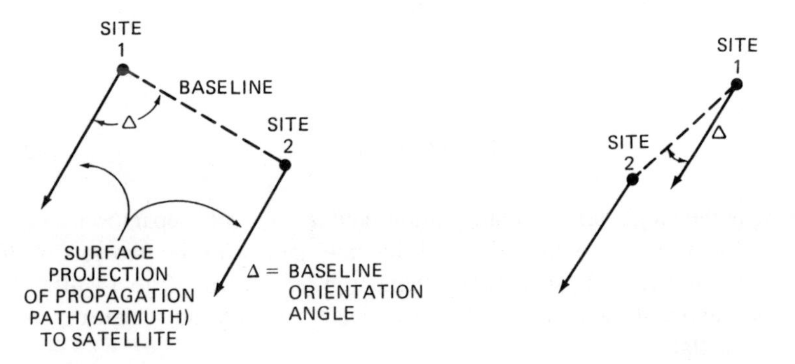

Figure 10-10. Baseline orientation in diversity systems.

Δ is small and both paths pass through the same volume in the troposphere, increasing the probability that a rain cell will intersect both paths much of the time.

The elevation angle to the satellite also impacts diversity performance because of the increased probability of intersecting a rain cell at low elevation angles. In general, the lower the elevation angle, the longer the site separation required to achieve a given level of diversity gain.

The operating frequency could also be expected to be a factor in diversity systems, since the probability of a given attenuation level being exceeded on a single path is heavily dependent on frequency. Diversity gain, however, is realized because of the physical configuration of the sites and the structure of rain cells, hence, at least to a first order, the frequency of operation would not play a major role in determining diversity gain for a given site configuration and weather region.

Just how these factors are *quantitatively* interrelated to diversity gain is difficult to assess analytically. An empirical analysis of an extensive set of diversity measurements was accomplished, however, by Hodge [10.5, 10.6]. His results have produced a comprehensive prediction model for diversity gain on earth–space propagation paths.

The original Hodge model [10.5] considered only the relationship between diversity gain and site separation. The model was based on a limited data base: 15.3 GHz measurements in Columbus, Ohio, and 16 GHz measurements in Holmdel, New Jersey. Site separations from 3 to 34 km were included. The resulting empirical relationship for diversity gain was

$$G_D = a(1 - e^{-bd}) \tag{10-3}$$

$$a = A_S - 3.6(1 - e^{-0.24A_S}) \tag{10-4}$$

and

$$b = 0.46(1 - e^{-0.26A_S}) \tag{10-5}$$

where A_S is the single terminal attenuation, and d is the site separation.

Figure 10-11 compares the predicted diversity gain with the two sets of diversity measurements for single site attenuation values from 2 to 14 dB. The prediction curves show a sharp knee at about 8 km, beyond which diversity gain remains flat.

The improved diversity gain model [10.6], utilized an expanded data base of 34 sets of measurements, and included the dependence of diversity gain on frequency, elevation angle, and baseline orientation, as well as site separation. The resulting empirical result for diversity gain was

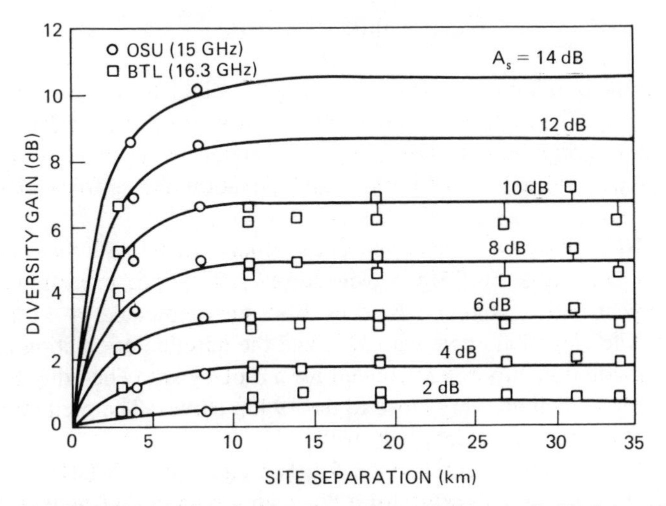

Figure 10-11. Empirical path diversity gain prediction, early Hodge model.

$$G_D = G(d, A_S)\ G(f)\ G(\theta)\ G(\Delta) \tag{10-6}$$

where

d = site separation, in km
A_S = single terminal attenuation, in dB
f = frequency, in GHz
θ = elevation angle, in degrees
Δ = baseline orientation angle, in degrees (as defined in Fig. 10-10).

Each of the gain functions were found to be

$$G(d, A_S) = a(1 - e^{-bd}) \tag{10-7}$$

with

$$a = 0.64A_S - 1.6(1 - e^{-0.11A_S}) \tag{10-8}$$

and

$$b = 0.585(1 - e^{-0.098A_S}) \tag{10-9}$$

$$G(f) = 1.64e^{-0.025f} \tag{10-10}$$

$$G(\theta) = 0.00492\theta + 0.834 \tag{10-11}$$

$$G(\Delta) = 0.00177\Delta + 0.887 \tag{10-12}$$

(Note that the a and b coefficients for $G(d, A_S)$ differ from those found in the earlier model.) The improved model had an r.m.s. error of 0.73 dB when compared with the original data sets, covering frequencies from 11.6 to 35 GHz, site separations from 1.7 to 46.9 km, and elevation angles from 11 to 55 degrees.

Figure 10-12 shows a comparison of the Hodge model with measured diversity gain data taken at 16 GHz in New Jersey [10.7]. The measurements were for a one year period, and consisted of three sites separated by 11.2, 19, and 30.4 km. The elevation angle was 32°, and the baseline orientation angle was 90°. The prediction curve is calculated for a d of 19 km (The other two values of d give curves that are very close to the 19 km curve). The prediction is seen to represent the measured data very well.

Figure 10-13 shows a comparison for data taken at 17.8 GHz in Colorado [10.8]. The measurements were for a nineteen month period, with a site separation of 33.3 km, an elevation angle of 42.6°, and a baseline orientation of 60°. The prediction curve is 0.5–1 dB below the measured diversity gain.

The prediction curve for the Etam, West Virginia measurements discussed in the previous section is shown in Figure 10-7. The prediction again tends to underpredict the measured diversity gain by about 1 dB for Lenox and by about 2 dB for Etam.

The variation of diversity gain with operating frequency can be seen on Fig-

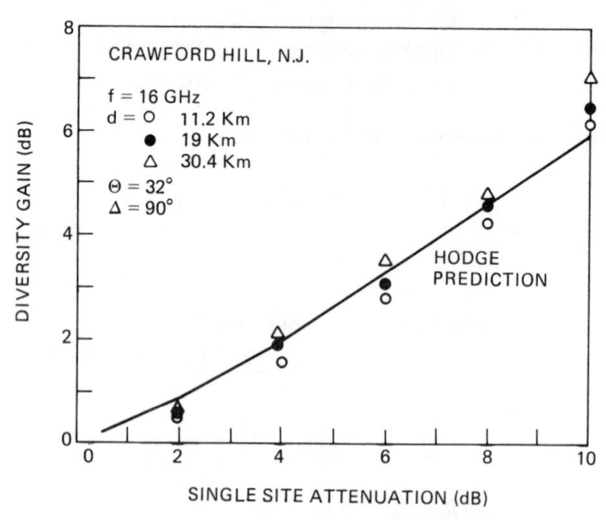

Figure 10-12. Comparison of Hodge prediction with divesity gain measurements in Crawford Hill, NJ.

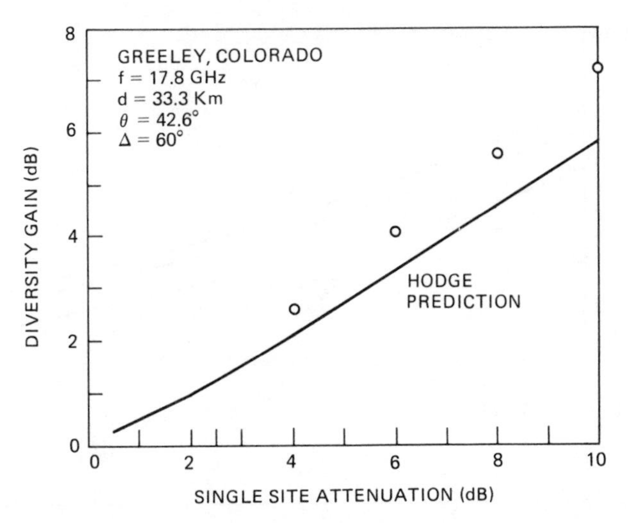

Figure 10-13. Comparison of Hodge prediction with diversity gain measurements in Colorado.

ure 10-14, which shows plots of the Hodge prediction model at 12, 20, and 30 GHz, for site separations out to 25 km. The single site attenuation was set to 10 dB, with the elevation angle at 35 degrees and the baseline orientation at 90 degrees.

Figure 10-15 shows the frequency dependence directly, for three specific values of site separation, with single site attenuation set to 10 dB.

The two previous figures tend to imply that diversity gain degrades with in-

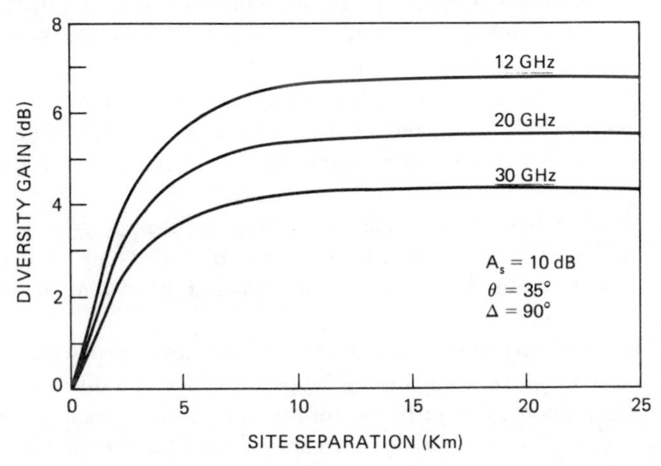

Figure 10-14. Diversity gain as a function of site separation and frequency.

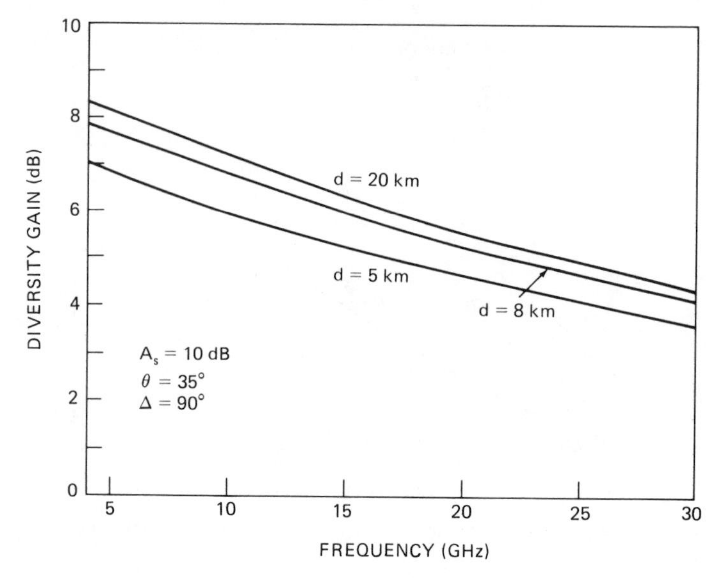

Figure 10-15. Frequency dependence of diversity gain for a fixed single site attenuation.

creasing frequency, however, that is not the case. The two sets of plots are for a fixed value of single site attenuation A_S, thus they do not represent the same conditions at each frequency. A 10 dB attenuation at 12 GHz corresponds to a much more severe weather condition than a 10 dB attenuation at 30 GHz. A better appreciation of the frequency dependence can be observed by fixing the *rain rate*, assuring that the same weather conditions exist for all frequencies. This situation is shown in Figure 10-16, for a rain rate of 25 mm/h. The single site attenuation for each frequency was calculated from the CCIR coefficients of Table 4-3, with an assumed storm height of 4 km.

A dramatic increase in diversity gain with frequency is observed throughout the range, with a slight reduction at frequencies above 30 GHz. Note that there is little difference between the site separation plots (5 and 20 km) shown on the figure.

Figure 10-17 and 10-18 show the dependence of diversity gain on elevation angle and on baseline orientation, respectively. Both plots are for a fixed rain rate of 25 mm/h and a site separation of 20 km. Little variation with either parameter is observed.

A useful set of diversity design curves can be developed with the Hodge model to allow first level estimates of the performance of a diversity link for a given set of conditions. Figure 10-14, for example, gives an estimate of the site separation required to achieve a given improvement for two station diversity. The curves show that two sites with 6 dB rain attenuation margins (i.e., a di-

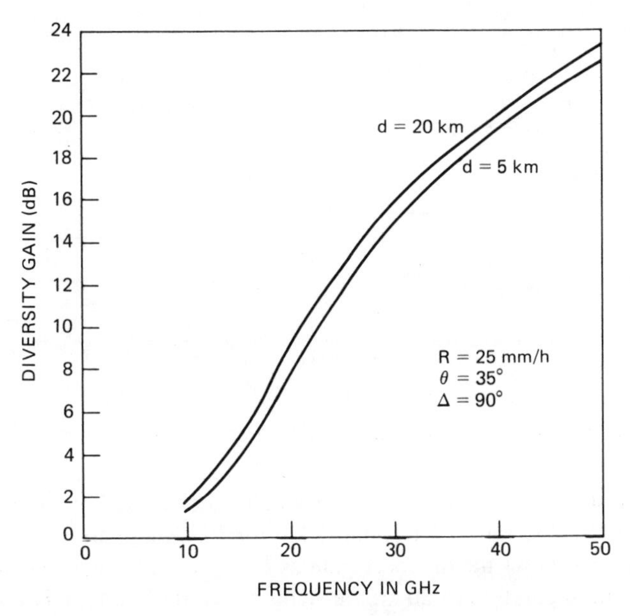

Figure 10-16. Frequency dependence of diversity gain for a fixed rain rate.

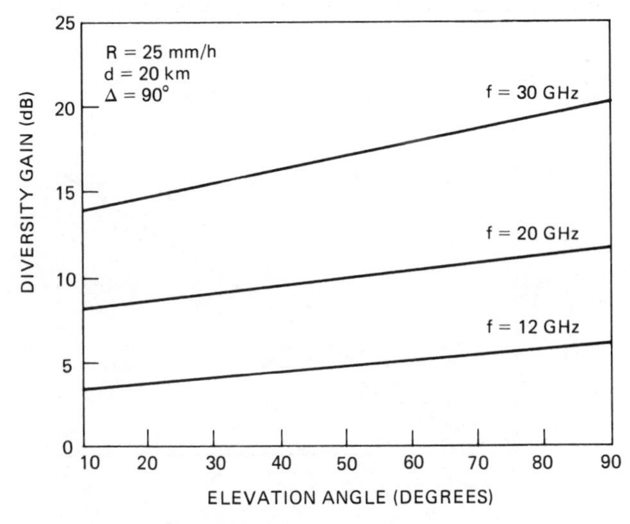

Figure 10-17. Diversity gain as a function of elevation angle.

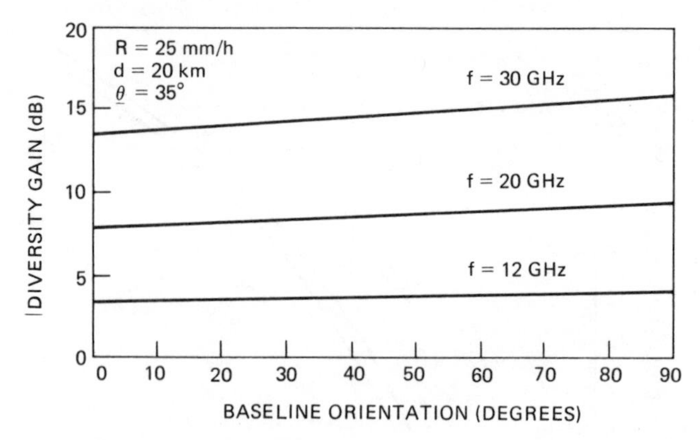

Figure 10-18. Diversity gain as a function of baseline orientation.

versity gain of 4 dB since A_S = 10 dB) should be separated by at least 2.5 km for 12 GHz, 3.5 km for 20 GHz, and 7 km for 30 GHz operation.

Figure 10-19 shows the diversity gain as a function of single site attenuation for a fixed site separation of 20 km, at frequencies of 12, 20, and 30 GHz. The diversity gain at a given percent of the year can be determined from the figure. For example, consider the case where a single site attenuation distribution at 12 GHz shows that a 15 dB power margin would be required to maintain a 99.99% availability. Reference to Figure 10-19 indicates that the implementation of two station diversity with a separation of 20 km could achieve that same availability with a power margin of 4.5 dB at each site.

The Hodge prediction model can also be used to generate joint (diversity)

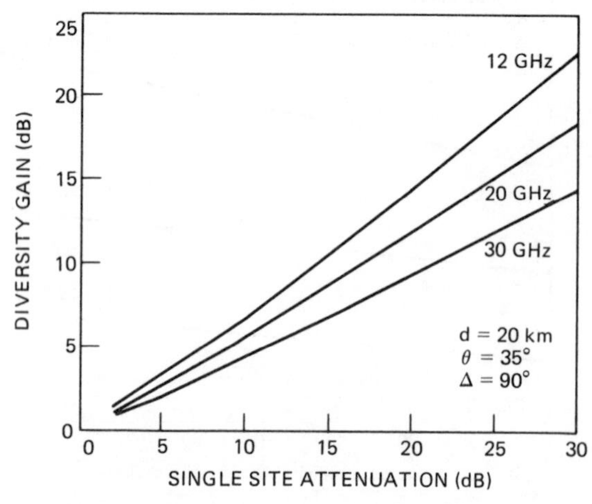

Figure 10-19. Diversity gain as a function of single site attenuation.

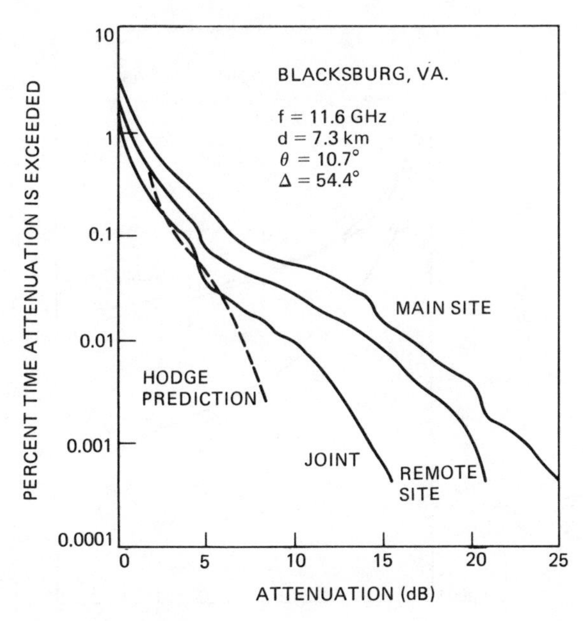

Figure 10-20. Joint attenuation prediction compared to measured data, Blacksburg, VA.

attenuation distributions from measured or predicted single site attenuation distributions. Figure 10-20 shows a plot of the Hodge prediction for the joint distribution, compared with the measured distributions, for the 11.6 GHz Virginia measurements described earlier (Figure 10-8). The prediction curve was developed from the single site attenuation curves, with the mean value of the two single site values used as the reference. The prediction is seen to agree fairly well to about the 0.02% level, however below that value the prediction gives much more optimistic results than that measured.

Figure 10-21 shows another example of the joint attenuation distribution predicted from measured single site attenuation. The measurements were taken in Austin, Texas with 13.6 GHz radiometers located 15.8 km apart [10.9]. The prediction shows excellent agreement with the measured distribution.

Figrue 10-22(a–d) shows four further examples of predicted joint distributions determined from long term attenuation measurements. The data, previously presented in Figure 4-5, was measured at 11.7 GHz with the CTS beacon. Prediction curves for a site separation of 20 km and a baseline orientation of 90 degrees are shown.

Other prediction models are available for the evaluation of diversity performance on earth–space links. Kaul [10.10] introduced meteorological considerations to the Hodge model and determined the maximum available diversity gain as a function of convective (thunderstorm) and stratiform (uniform) rain.

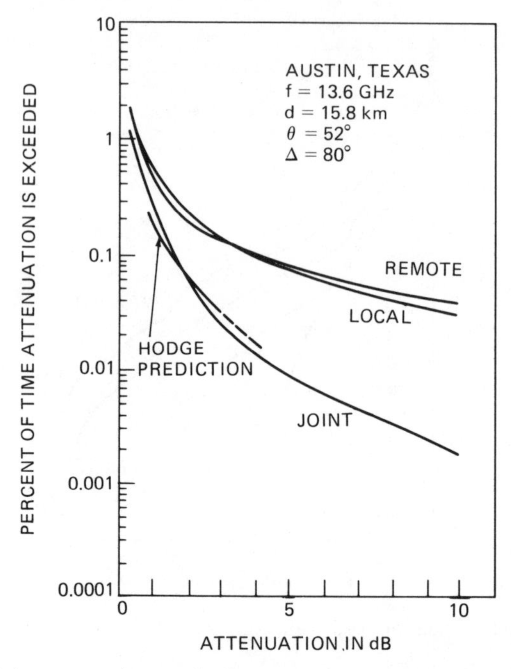

Figure 10-21. Joint attenuation prediction compared to measured data, Austin, TX.

The concept of relative diversity gain was developed by Goldhirsh and radar measurements were used to construct diversity statistics for various single site attenuation and site separations [10.11].

Modifications and enhancements to site diversity prediction procedures can be expected as more measured diversity attenuation data and operational site diversity statistics become available.

10.1.4. Diversity Processing

The selection of which site is to be on-line in a site diversity operation can be made according to several criteria. The ultimate objective is to select the site with the lowest rain attenuation (uplink diversity) or the site with the highest received signal level (downlink diversity). Unfortunately, it is not always possible to use these simple criteria fully, because of practical implementation problems.

For the uplink case, the attenuation level can be determined directly by monitoring a beacon transmitted from the satellite, or indirectly from radiometer or radar measurements. Alternatively, the received signal level at the satellite could be telemetered down to the ground, or, if the site also employs downlink di-

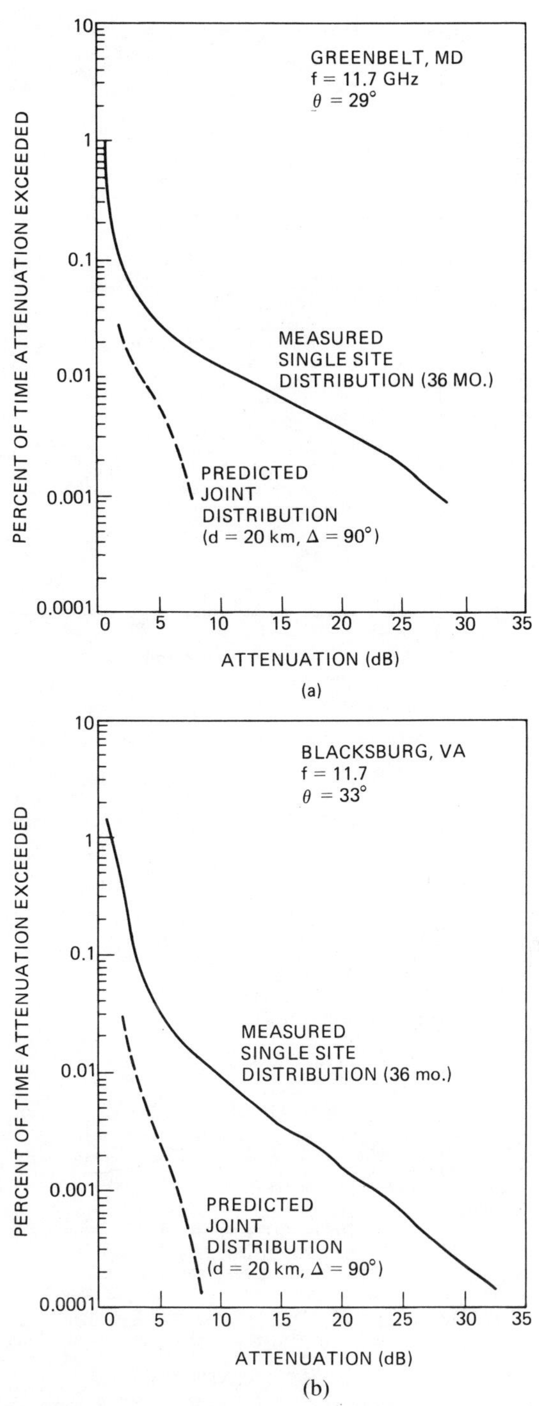

(b)

Figure 10-22. Diversity prediction from measured long term attenuation distributions.

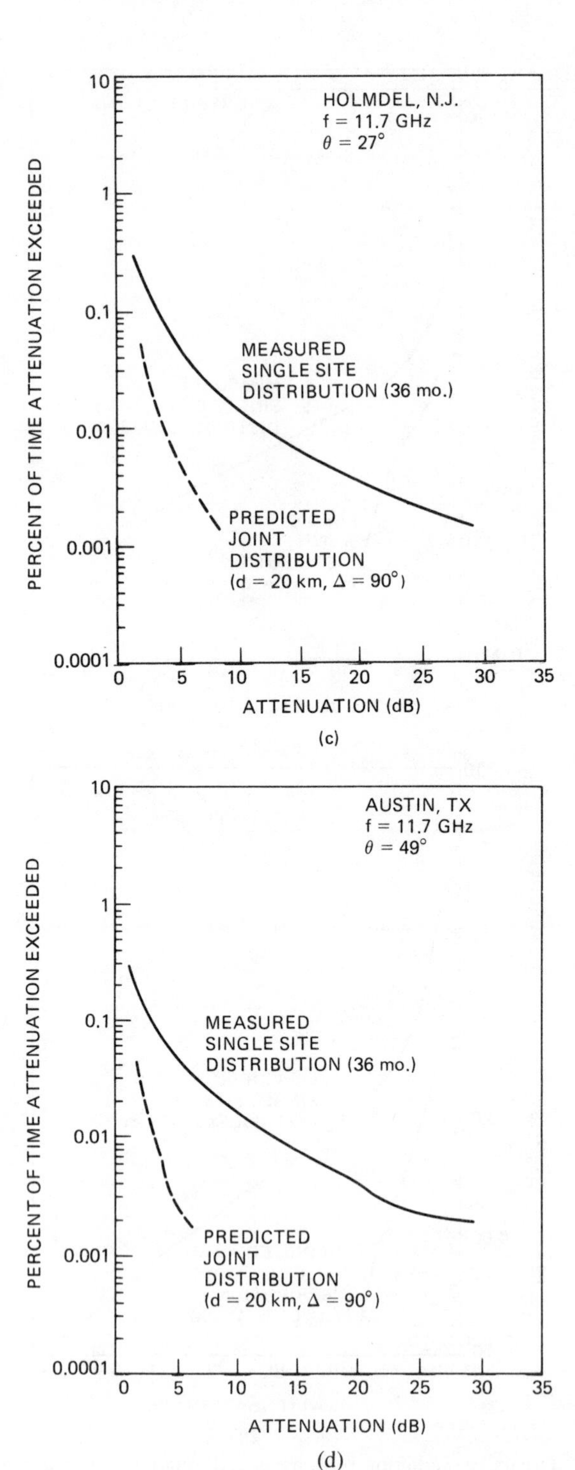

(c)

(d)

Figure 10-22. *(Continued)*

versity, the downlink decision process could be applied to the uplink as well. Uplink diversity is much more difficult to implement than is downlink diversity, since the switching of high power signals is involved, and since the actual attenuation on the link may not be known accurately or quickly enough to allow for a ''safe'' switchover. Data could be lost during the switchover, particularly for analog transmissions. Uplink diversity with digital signals is less restrictive; however, accurate timing and delay information must be maintained between the uplink sites and the origination location of the uplink information signal.

For the downlink case, the two received signal levels can be compared and the highest level signal selected for ''on-line'' use. Several alternative decision algorithms can be implemented:

Primary Predominant. A primary site can be on-line as long as its signal level remains above a preset threshold. The secondary site is only brought on-line if the primary site is below threshold and the secondary site is above threshold.

Dual Active. Both sites (or three in the case of three site diversity) are active at all times, and the diversity processor selects the highest level signal for further processing.

Combining. If the signal format allows it, the signals from each site could be additively combined and the combined signal used as the on-line input. No switching would be required. This techique is particularly attractive for analog video or voice transmissions.

In any diversity processing system, particularly those involving digitally formatted signals, or burst formatted signals such as time division multiple access (TDMA), synchronization and timing are critical for successful operation.

An experimental diversity system utilizing an INTELSAT V satellite and TDMA/DSI (digital speech interpolation), investigated methods and techniques for TDMA site diversity operation in Japan at K_U band [10.12]. A short diversity burst is provided in the burst time plan to maintain burst synchronization for diversity operation (see Figure 10-23). The separation distance was 97 km, and the elevation angles were 6.8° for the prime site (Yamaguche) and 6.3° for the backup site (Hamada).

A threshold of signal quality was defined in terms of the bit error rate (BER) from the following algorithm,

$$Q_{th} = \frac{N_S}{F_S N_0 T_S} \tag{10-13}$$

where

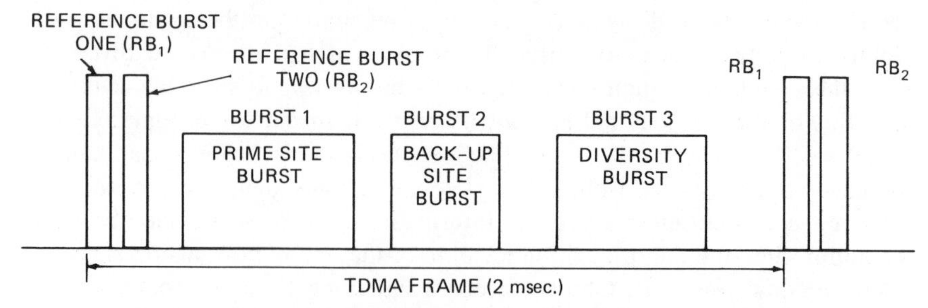

Figure 10-23. Burst time plan for TDMA diversity experiment.

Q_{th} = threshold of signal quality

N_S = number of error bits specifying the threshold

F_S = TDMA frame frequency (500 Hz)

N_0 = number of data bits observed for error detection in each frame

T_S = period of time for error detection

For the experiment N_S was set to 100, N_0 to 10,000, and T_S to 20 seconds, corresponding to a 1×10^{-6} BER.

Figure 10-24 shows the diversity switching response observed for a rain event of 50 minutes duration. Figure 10-25 shows the cumulative distribution of the bit error rate for the same event. Distribution curves are shown for the prime site burst, the backup site burst, and the diversity burst. Improvement in BER performance can be seen by employment of diversity operation.

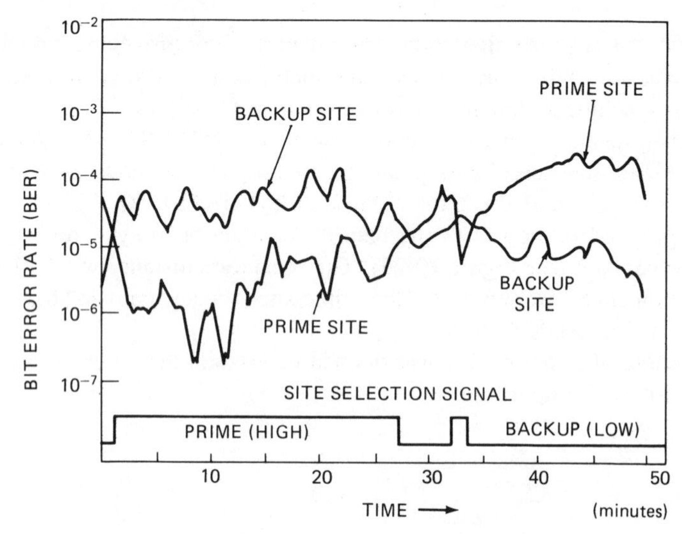

Figure 10-24. Diversity switching response for TDMA diversity experiment.

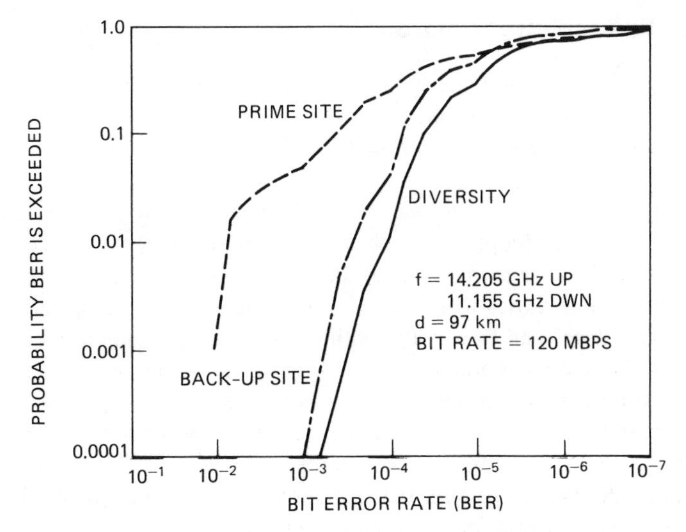

Figure 10-25. Cumulative distribution of bit error rate for looped back and diversity operation.

10.2. POWER CONTROL

Power control refers to the process of varying transmit power on a satellite link, in the presence of rain attenuation, to maintain a desired power level at the receiver. Power control attempts to restore the link by increasing the transmit power during a rain attenuation event, then reducing power after the event back to its clear sky value.

Generally, it is desirable to keep the transmit power at the minimum acceptable level necessary to overcome the rain attenuation, in order to avoid interference to other satellites (in the case of uplink power control) or other ground terminals (in the case of downlink power control).

Power control requires a knowledge of the path attenuation on the link to be controlled. The particular method of obtaining this information depends on whether uplink or downlink power control is being implemented, and in the particular configuration of the space communications system.

Power control is generally only applicable to single service, single user type links, since large carrier level variations on multi-carrier, multi-user systems could cause significant interference between carriers operating through the same transponder.

Considerations relating to the unique characteristics of uplink and downlink power control are discussed in the following sections.

10.2.1. Uplink Power Control

Uplink power control provides a direct means of restoring the uplink signal loss during a rain attenuation event. It is utilized in fixed satellite service applications and has been considered for broadcast satellite service feeder links (uplinks) also.

Two types of power control can be implemented, *closed loop* or *open loop* systems. In a closed loop system, the transmit power level is adjusted directly as the detected receive signal level at the satellite, returned via a telemetry link back to the ground, varies with time. Control ranges of up to 20 dB are possible, and response times can be nearly continuous if the telemetered receive signal level is available on a continuous basis. Figure 10-26 shows a simplified block diagram of the closed loop uplink power control system.

In an open loop power control system, the transmit power level is adjusted by operation on a radio frequency control signal which itself undergoes rain attenuation, and is used to infer the rain attenuation experienced on the uplink. The radio frequency control signal can be either:

(a) the downlink signal,
(b) a beacon signal at or near the uplink frequency, or
(c) a ground based radiometer or radar.

Figure 10-27(a-c) shows block diagrams for the three open loop techniques. Broadcast satellite service frequencies, 17 GHz uplink and 12 GHz downlink, are used for the examples shown on the figure.

In the downlink control signal system, Fig. 10-27(a), the signal level of the 12 GHz downlink is continuously monitored and used to develop the control signal for the high power transmitter. The transmitter may be either a solid state (SS) or a traveling wave tube amplifier (TWTA). The control signal level is determined in the processor from rain attenuation prediction models which compute the expected uplink attenuation at 17 GHz from the measured downlink attenuation at 12 GHz.* The downlink control signal method is the most prevalent type of uplink power control, because of the availability of the downlink at the ground station and the relative ease of implementation.

In the beacon control signal system, Fig. 10-27(b), a satellite beacon signal, preferably in the same frequency band as the uplink, is used to monitor the rain attenuation in the link. The detected beacon signal level is then used to develop the control signal. Since the measured signal attenuation is at (or very close to) the frequency to be controlled, no estimation is required in the processor. This method provides the most precise power control of the three techiques.

**Note:* This method requires a determination of the *attenuation ratio* between the uplink and the downlink. The attenuation ratio can change very quickly with time, since it is heavily dependent on drop size distribution and rain rate along the path.

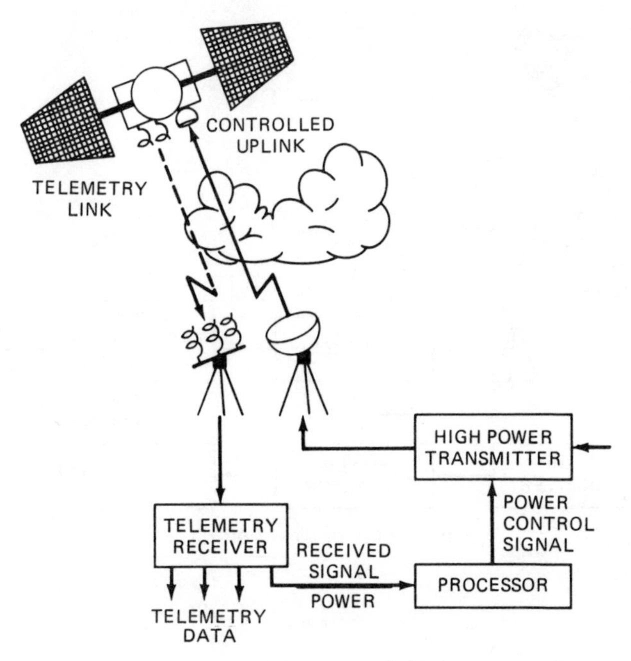

Figure 10-26. Closed loop uplink power control.

The third method, Fig. 10-27(c), develops an estimation of the uplink rain attenuation from sky temperature measurements with a radiometer directed at the same satellite path as the uplink signal.

Rain attenuation can be estimated directly from the noise temperature T_S, as discussed in Section 7.3 and shown in Figure 7-9. The attenuation A(dB) can be determined from T_S by inverting Equation (7-7), i.e.,

$$A(\text{dB}) = 10 \log_{10} \left[\frac{T_m}{T_m - T_S} \right] \tag{10-14}$$

where T_S is the mean path temperature, in °K. The variation in attenuation with path temperature can produce about a 1 dB error in the prediction at high values of attenuation, as shown in Figure 7-9. This power control method is the least accurate of the three described, and is generally only implemented if no other means of determining the path attenuation are available.

Uplink power control does have some basic limitations, no matter which method is employed. Often it is difficult to maintain the desired power flux density (PFD) at the satellite to a reasonable accuracy, say ±1 dB, because of (a) measurement errors in the detection or processing of the control signal, (b)

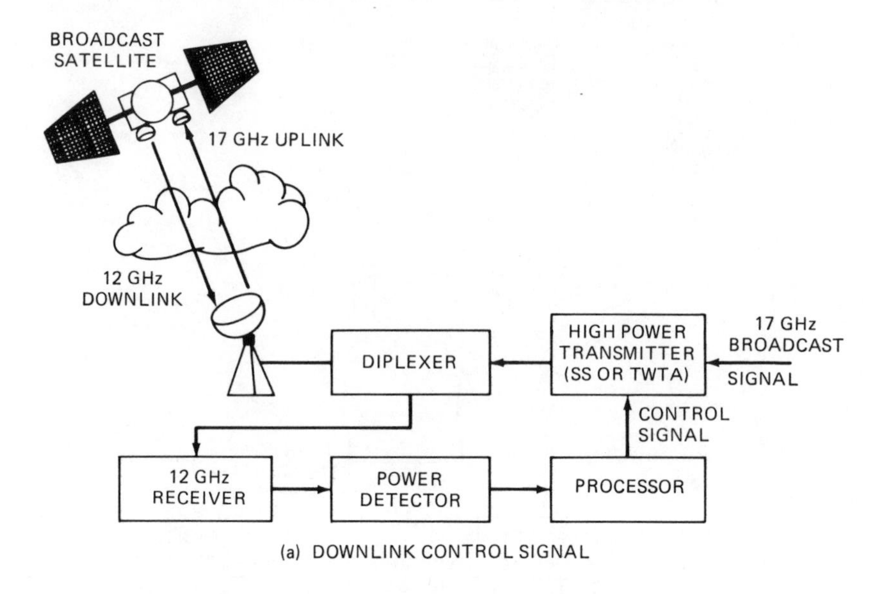

(a) DOWNLINK CONTROL SIGNAL

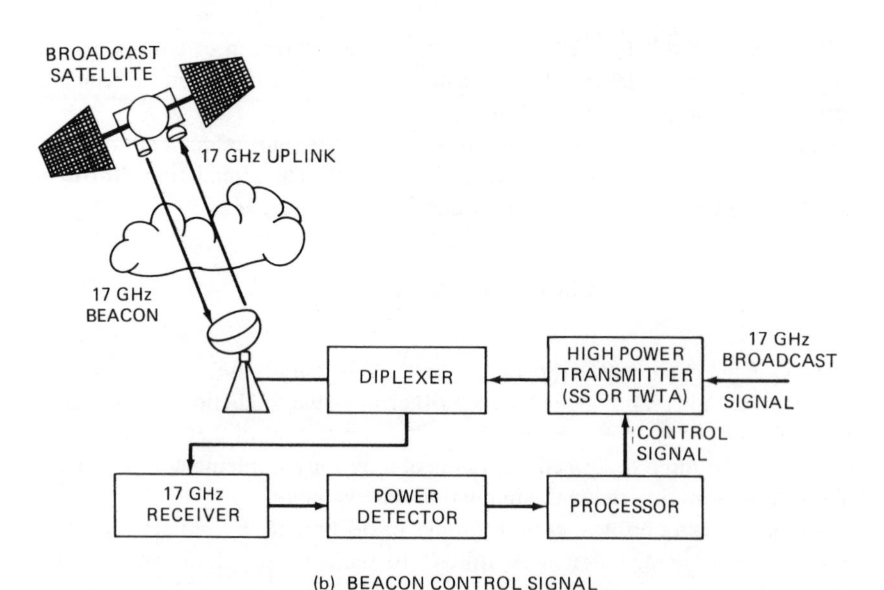

(b) BEACON CONTROL SIGNAL

Figure 10-27. Open loop uplink power control techniques.

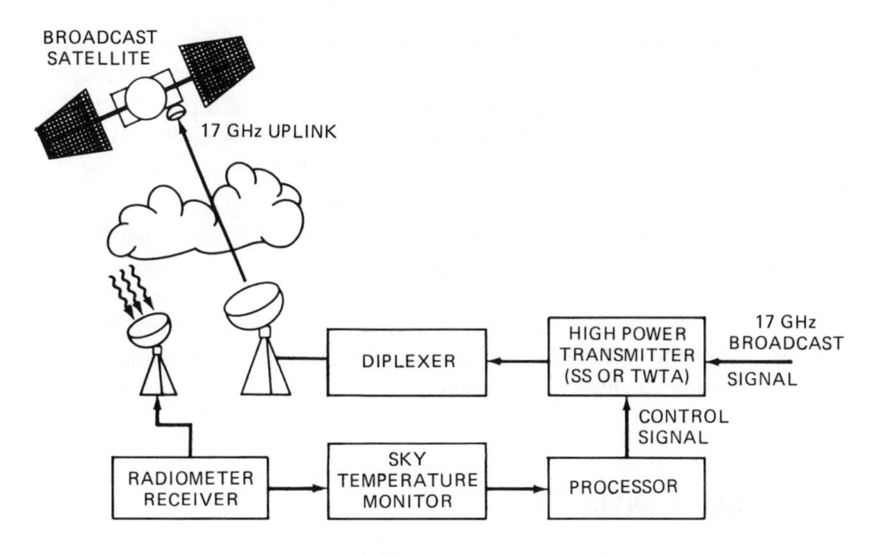

(c) GROUND BASED RADIOMETER CONTROL SIGNAL

Figure 10-27. *(Continued)*

time delays due to the control operation, or (c) uncertainty in the prediction models used to develop the uplink attenuation estimations.

Also, attenuation in intense rain storms can reach rates of 1 dB per second. Rates of this level are difficult to compensate for completely because of response times in the control system.

Power control can induce added interference due to *cross polarization* effects in rain. Since the cross-polarized component increases during rain, an increase in the transmit power will also increase the cross polarized component, increasing the probability of interference to adjacent cross-polarized satellites in nearby orbital positions.

Uplink power control measurements have been reported using 14/11 GHz TDMA links with INTELSAT V [10.12], and 14/12 GHz FM-TV links with the BSE satellite [10.13]. Both experiments utilized open loop power control with a downlink frequency control signal. Reference 10.12 reported excellent performance for uplink attenuation due to both rain and to scintillations. Reference 10.13 reported that the variation of uplink receiving power due to rain was kept to within 0.5 dB R.M.S. and 1.5 dB peak-to-peak, most of the time.

10.2.2. Downlink Power Control

Power control on a satellite downlink is generally limited to one or two fixed level switchable modes of operation to accomodate rain attenuation losses. The

NASA ACTS (Advanced Communications Technology Satellite), for example, which operates in the 30/20 GHz frequency bands, has two modes of downlink operation [10.14]. The low power mode operates with 8 watts of RF transmit power, and the high power mode operates with 40 watts. A multimode TWTA is used to generate the two power levels. The high power mode provides about 7 dB of additional margin for rain attenuation compensation.

Downlink power control is not efficient in directing the additional power to a ground terminal (or terminals) undergoing a rain attenuation event, since the entire antenna footprint receives the added power. A satellite transmitter providing service to a large number of geographically independent ground terminals would have to operate at or near its peak power almost continuously in order to overcome the highest attenuation experienced by just one of the ground terminals.

10.3. ORBITAL DIVERSITY

Orbital diversity refers to the use of two widely dispersed on-orbit satellites to provide seperate converging paths to a single ground terminal. Diversity gain is realized by utilizing the link with the lowest path attenuation. Statistics similar in concept to site diversity operation can be generated.

Since orbital diversity requires two widely separated on-orbit satellites, its application is very limited. Also, orbital diversity requires two antenna systems at the ground terminal to be fully effective.

Orbital diversity improvement is not primarily due to the celular structure of heavy rain, as for site diversity, but occurs because there will always be some amount of statistical decorrelation between two separate paths to a single ground terminal experiencing rain.

Figure 10-28 shows two examples of orbital diversity parameters for two COMSTAR satellites with a terminal in Washington, DC, and for two INTELSAT satellites with a terminal in San Francisco.

The COMSTAR satellites are separated by 33° in orbit, providing two links with elevation angles at 21.4° and 41.3°. The INTELSAT satellite pair is separated by 126°, and the two resulting links to San Francisco are widely separated in azimuth angle.

The degree of improvement obtained by orbital diversity over single site operation would not be expected to be as great as site diversity operation because both paths converge at the lower end to the same point on the earth. Since most of the rain attenuation occurs in the lower 4 km of the troposphere, there would be little statistical independence between the two diverging paths.

Orbital diversity measurements have shown this to be the case. Figure 10-29 shows measured orbital diversity data taken at Palmetto, GA, using the 19 GHz COMSTAR D2 beacon and a radiometer pointed in the direction of COMSTAR

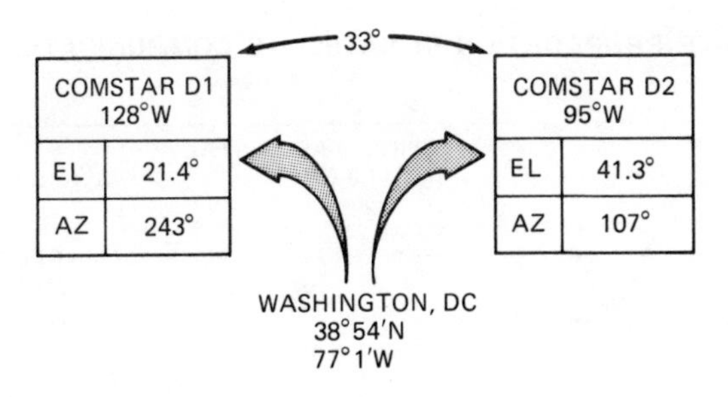

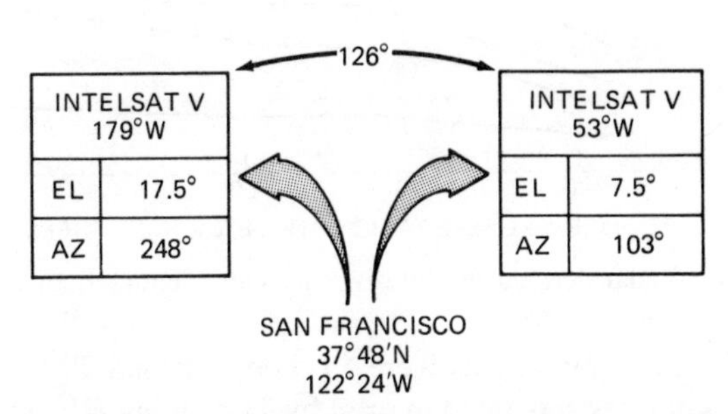

Figure 10-28. Examples of orbital diversity characteristics.

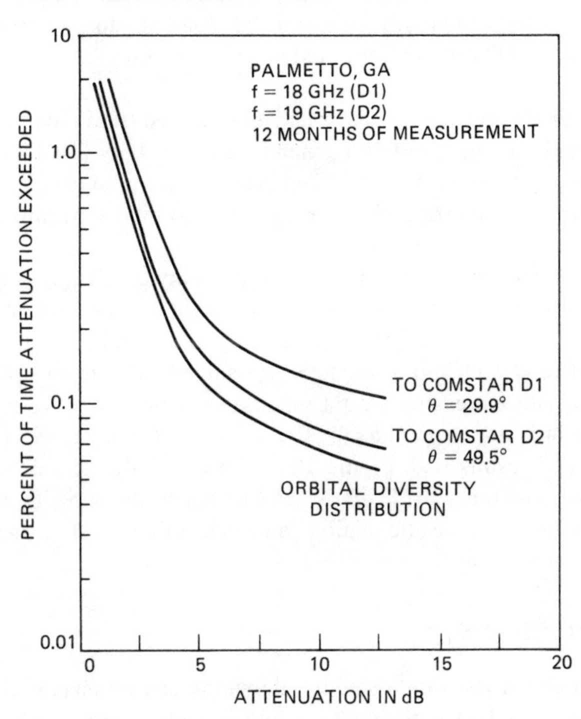

Figure 10-29. Orbital diversity measurements at Palmetto, GA.

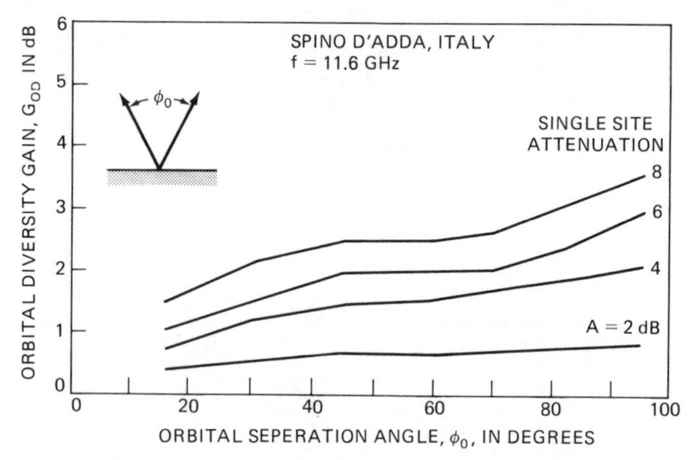

Figure 10-30. Radar derived orbital diversity gain characteristics at 11.6 GHz.

D1 [10.15]. The elevation angles were 49.5 degrees and 29.9 degrees respectively, with the satellites separated in orbit by 41 degrees. The results show that rain attenuation on the two paths was highly correlated, and that the orbital diversity gain is low. Diversity gain at an attenuation value of 10 dB was 2 dB (referenced to the D2 distribution), and at an attenuation value of 5 dB was less than 1 dB.

Orbital diversity and site diversity statistics derived from radar measurements were developed and analyzed at a ground station in Italy [10.16]. Attenuation measurements with the 11.6 GHz SIRIO beacon were used to calibrate the radar measurements. Various paths were evaluated for orbital separations from 16 to 95 degrees.

Figure 10-30 shows a plot of the orbital diversity gain G_{oD}, as a function of the satellite orbital separation angle, ϕ_o, for single site attenuation values from 2 to 8 dB. The curves are scaled to a frequency of 11.6 GHz, and a symmetrical configuration (equal elevation angles) is assumed. The results show a slight improvement with increasing orbital separation angle.

Improvement is not as good as for site diversity, however, as can be seen by comparing these results with Figure 10-11 for the Hodge site diversity model. Site diversity gain for $A = 8$ dB is seen to reach about 5 dB, while orbital diversity gain for the same attenuation value is less than 4 dB, even at extremely large orbital separations.

10.4. SPOT BEAMS

The received signal level on a satellite downlink can be increased during path attenuation periods by switching to a higher gain satellite antenna. The increased antenna gain, corresponding to a narrower antenna beam, concentrates the power into a smaller area on the earth's surface, resulting in a higher EIRP at the ground terminal undergoing attenuation. A satellite serving many ground terminals may have one or more of these *spot beams* available to switch over to during rain attenuation events. The spot beams could be directed to the desired ground terminal location either by mechanical movement of a separate reflector or by an electronically switched antenna feed system.

Figure 10-31(a)–(d), shows coverage area contours (footprints) for four typical antenna beam types as viewed from a geosynchronous satellite located over the United States. The 3 dB (half-power) beamwidth and on-axis gain (assuming 55% efficiency) are shown with each contour. CONUS coverage is typical for fixed satellite service systems, while time zone beams are useful for direct broadcast applications. Regional and metropolitan area spot beams are used in frequency reuse systems as well as for rain attenuation restoration applications.

The increase in EIRP at the ground terminal can be very significant, as displayed in Figure 10-32, which shows the dB improvement available between each of the beam coverage types. For example, the use of a metropolitan area spot beam in place of a CONUS coverage beam would provide 24.1 dB of additional EIRP. Switching from a time zone beam to a reginal spot beam gives 5.4 dB of additional power.

Table 10-1 shows the antenna diameter required on the satellite for each of the coverage types, for frequencies of 4, 12, 20, and 40 GHz, assuming again a 55% antenna efficiency.

10.5. SIGNAL PROCESSING RESTORATION TECHNIQUES

All of the methods discussed previously in this chapter for overcoming severe path attenuation have not altered the basic signal format in the process of restoring the link. Site diversity and orbital diversity involve selective switching between two or more redundent links bearing the same information signal. Power control and spot beams involve and increase in the signal power or the EIRP to "burn through" a severe attenuation event. Data rates and information rates are not altered, and no processing is required on the signal format itself.

There are restoration techniques, however, that can be implemented by modifying the basic characteristics of the signal. These characteristics include carrier frequency, bandwidth, data rate, and coding scheme. Signal modification techniques tend to be more complex and potentially more costly, since extensive signal processing, both on-board the satellite and on the ground, is required. Several signal processing restoration techniques are discussed in this concluding section on techniques for overcoming severe attenuation.

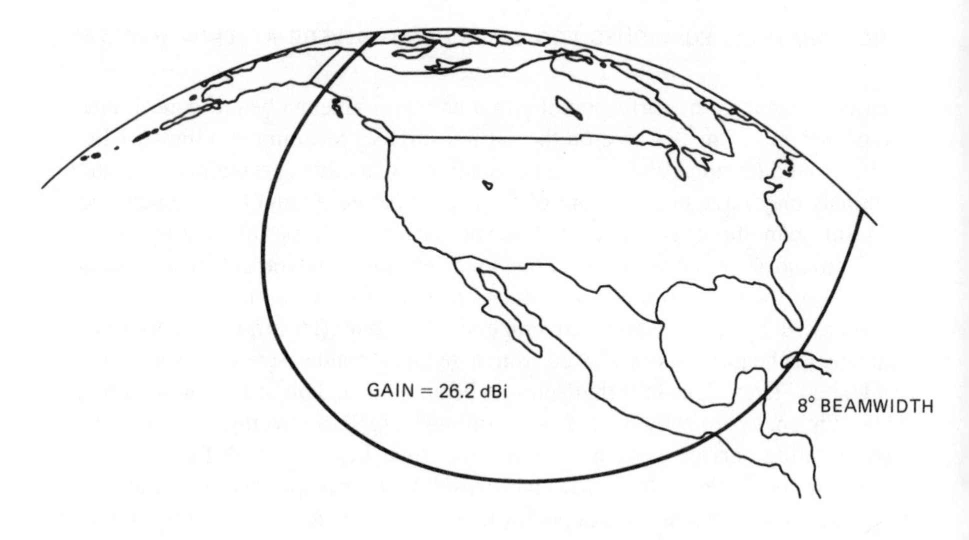

(a) CONTINENTAL U.S. (CONUS) BEAM

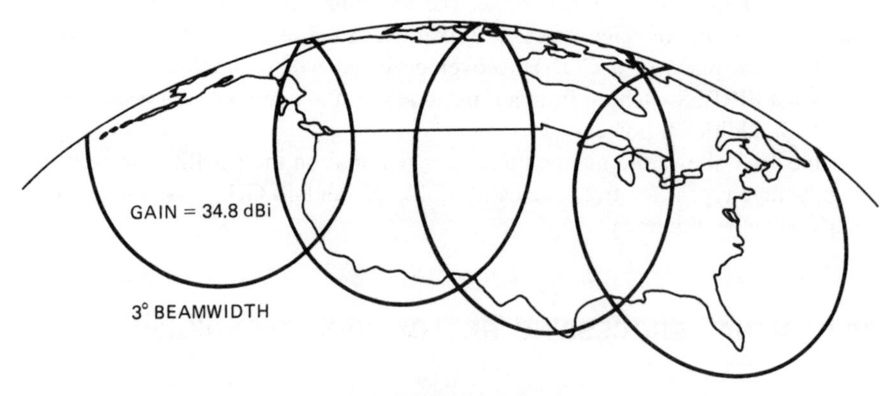

(b) TIME ZONE BEAMS

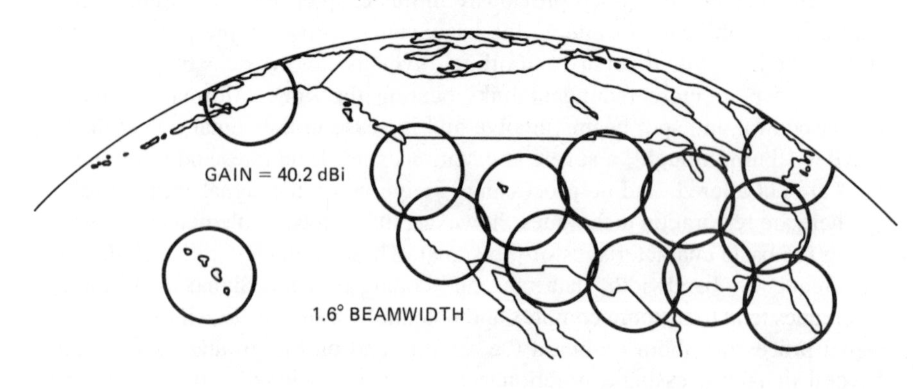

(c) REGIONAL SPOT BEAMS

Figure 10-31. Antenna beam coverage contours.

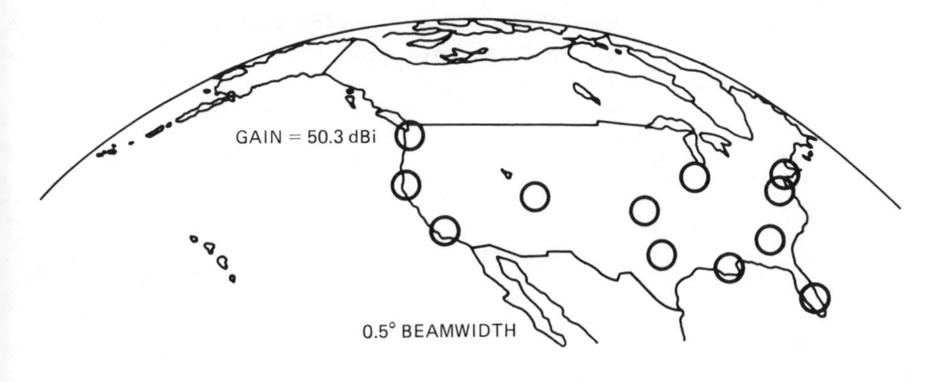

(d) METROPOLITAN AREA SPOT BEAMS

Figure 10-31. *(Continued)*

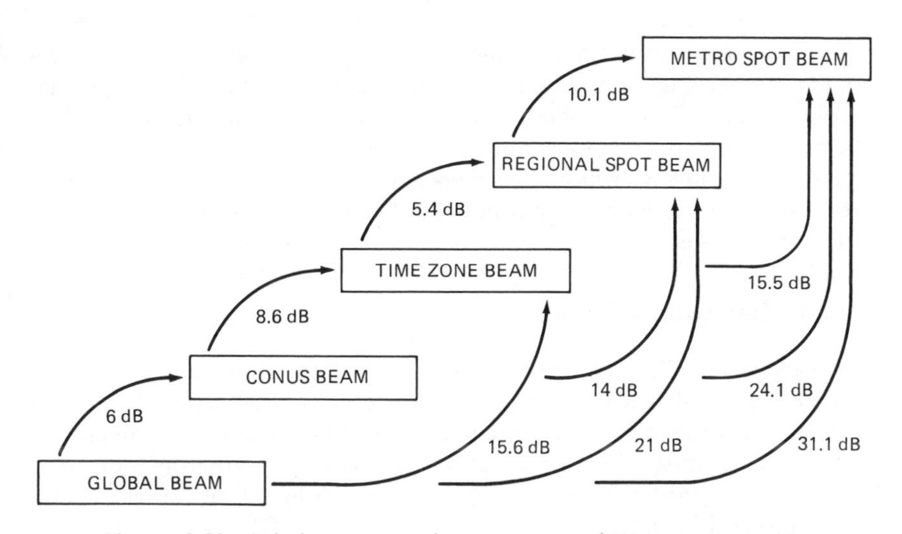

Figure 10-32. Gain improvement between antenna beam coverage types.

10.5.1. Frequency Diversity

In this technique, earth–space links operating in frequency bands subject to rain attenuation, e.g., 14/12 GHz or 30/20 GHz, are switched to a lower frequency band, such as 6/4 GHz, where rain attenuation is negligible, whenever a specified margin is exceeded. Ground stations and satellites employing this method must be equiped for dual frequency operation.

The number of lower frequency transponders required for a given satellite depends on the number of links being served simultaneously by the satellite,

Table 10-1. Satellite Antenna Diameters for Various Antenna Beam Coverage Types.

Coverage Type (Beamwidth)	Antenna Gain (dBi)	Antenna Diameter (meters)			
		4 GHz	12 GHz	20 GHz	40 GHz
Global (18°)	19.2	0.29	0.096	0.06	0.03
Conus (8°)	26.2	0.65	0.22	0.18	0.09
Time zone (3°)	34.8	1.75	0.6	0.33	0.17
Regional spot (1.6°)	40.2	3.3	1.1	0.66	0.33
Metro area spot (0.5°)	50.3	10.5	3.5	2.1	1.05

and on the probability of rain attenuation exceeding the rain margin for any one link. For example, assume a system with one link per transponder, and a 10% probability that rain attenuation is exceeded on any one link. The number of lower frequency transponders required to maintain a 99.99% availability for a 48 transponder satellite would be 10, while the number for a 24 transponder would be 8 [10.17]. If the probability of rain on any one link is reduced to 1%, the number of lower frequency transponders drops to 3 and to 2, respectively.

For hybrid satellites which already operate in two or more frequency bands, frequency diversity may be a practical and low cost method of restoration for many applications.

10.5.2. Bandwidth Reduction

The bandwidth of the information bearing signal on either the uplink or the downhill can be reduced during periods of intense attenuation, resulting in an increase in available carrier-to-noise ratio on the link. Equations (9-16) and (9-17) show the inverse relationship between bandwidth and carrier-to-noise ratio. A reduction in bandwidth by one-half would result in a 3 dB carrier-to-noise improvement for that link.

Bandwidth reduction is obviously limited to those applications where a change in information rate or data rate can be tolerated. It is more easily implemented in digital systems, and in links where signal adaption delays are acceptable.

10.5.3. Transmission Delay

Another useful restoration technique which can be implemented when real-time operation is not essential (such as in bulk data transfer, electronic mail, etc.) involves temporarily storing the data during the rain attenuation period, and transmitting after the event has ended. Storage periods of several minutes to several hours would be required, depending on the frequency of operation and the rain conditions at the ground station in the system.

10.5.4. Adaptive Forward Error Correction

Source encoding is very useful for reducing the bit error rate, BER, on digital communications links, particularly those involving time division multiple access (TDMA) architectures. Coding gain of up to 8 dB can be achieved with some *forward error correction* (FEC) coding schemes. FEC improvement in BER is achieved by a reduction in sampling rate or in throughput on the link, however.

On TDMA links subject to rain attenuation or other degradation effects, *adaptive* FEC can provide a relatively efficient way to restore link availability during attenuation periods.

Adaptive FEC can be implemented in several ways on TDMA links. Generally, a small amount of the communications link capacity is held in reserve and is allocated, as needed, for *additional* coding to those links experiencing attenuation. The link data rate remains constant, since the extra capacity accommodates the additional coding bits needed for increased coding.

The excess capacity can be included at the TDMA *burst* level, or at the TDMA *frame* level. Adaptive FEC implementation at the burst level is only effective for downlink attenuation, since only individual bursts can be encoded. Frame level adaptive FEC can accomodate both uplink or downlink attenuation, since the reserve capacity can be applied to the entire frame for uplink restoration, or to portions of the frame (individual bursts) for downlink restoration.

Total adaptive FEC coding gains of up to 8 dB, for TDMA networks with 32 ground terminals, and operating at 14/11 GHz, have been reported [10.18]. The Advanced Communications Technology Satellite, ACTS, employs adaptive FEC and burst rate reduction, with on-board processing, to accommodate up to 10 dB of rain attenuation in the baseband processor mode [10.14].

10.6. SUMMARY

Many restoration techniques are available to reduce the effects of propagation impairments on space communications links. Each technique has its own advantages and disadvantages, both in implementation and in the level of attenuation which can be compensated for. Site diversity, for example, can provide well over 15 dB of restoration during intense rain attenuation events, while bandwidth reduction is limited to only a few dB because of signal parameter constraints.

The system designer can combine these techniques in many operational applications, to realize even further improvements. Uplink power control, together with downlink site diversity and adaptive FEC, could provide a significant restoration capability for some applications. Spot beams may provide acceptable levels of improvement if only one or two ground terminals are located in regions of intense rain.

Restoration techniques have provided a significant boost to the extension of satellite communications to higher frequency bands. Rain attenuation or other impairments, which can now be designed out of a system to a large extent, no longer present the barriers that were perceived to exist in earlier evaluations of the applicability of the higher frequency bands for satellite communications.

REFERENCES

10.1. Hodge, D. B., ''The Characteristics of Millimeter Wavelength Satellite-to-Ground Space Diversity Links,'' IEE Conference Publication No. 98, *Propagation of Radio Waves at Frequencies Above 10 GHz*, 10-13 April 1973, London, pp. 28–32.

10.2. Rogers, D. V., and Hodge, G., ''Diversity Measurements of 11.6 GHz Rain Attenuation at Etam and Lenox, West Virginia,'' *COMSAT Technical Review*, Vol. 9, No. 1, Spring 1979, pp. 243–254.

10.3. Towner, G. C., et al., ''Initial Results from the VPI&SU SIRIO Diversity Experiment,'' *Radio Science*, Vol. 17, No. 6, pp. 1489–1494, Nov.-Dec. 1982.

10.4. Tang, D. D., and Davidson, D., ''Diversity Reception of COMSTAR Satellite 19/29 GHz Beacons with the Tampa Triad, 1978-1981,'' *Radio Science*, Vol. 17, No. 6, pp. 1477–1488, Nov.-Dec. 1982.

10.5. Hodge, D. B., ''An Empirical Relationship for Path Diversity Gain,'' *IEEE Trans. on Antennas and Propagation*, Vol. 24, No. 3, pp. 250–251, March 1976.

10.6. Hodge, D. B., ''An Improved Model for Diversity Gain on Earth–Space Propagation Paths,'' *Radio Science*, Vol. 17, No. 6, pp. 1393–1399, Nov.-Dec. 1982.

10.7. Wilson, R. W., and W. L. Mammel, ''Results from a Three Radiometer Path Diversity Experiment,'' *Proceedings of Conf. on Propagation of Radio Waves at Frequencies Above 10 GHz*, Inst. of Electrical Engineers, London, pp. 23–27, April 1973.

10.8. Bergmann, H. J., ''Satellite Site Diversity: Results of a Radiometer Experiment at 13 and 18 GHz,'' *IEEE Trans. on Antennas and Propagation*, Vol. 25, No. 64, pp. 483–489, May 1977.

10.9. Vogel, W. J., ''Measurements of Satellite Beacon Attenuation at 11.7, 19.04, and 28.56 GHz and Radiometer Site Diversity at 13.6 GHz,'' *Radio Science*, Vol. 17, No. 6, pp. 1511–1520, Nov.-Dec. 1982.

10.10. Kaul, R., ''Extension of an Empirical Site Diversity Relation to Varying Rain Regions and Frequencies,'' URSI Symposium on Effects of the Lower Atmosphere on Radio Propagation Above 1 GHz, Lennoxville, Canada, pp. 4.13.1–1.13.4, May 26–30, 1980.

10.11. Goldhirsh, J., ''Sapce Diversity Performance Prediction for Earth–Satellite Paths using Radar Modeling Techniques,'' *Radio Science*, Vol. 17, No. 6, pp. 1400–1410, Nov.-Dec. 1982.

10.12. Watanabe, T. et al., ''Site Diversity and Up-Path Power Control Experiments for TDMA Satellite Link in 14/11 GHz Bands,'' Sixth International Conference on Digital Satellite Communications, Phoenix, Az., IEEE Cat. No. 83CH1848-1, pp. IX-21 to 28, 19–23, Sept. 1983.

10.13. Yamamoto, M. et al., ''Uplink Power Control Experiment,'' *IEEE Trans. on Broadcasting*, Vol. BC-28, No. 4, pp. 157–159, Dec. 1982.

10.14. Holmes, W. M. Jr., and G. A. Beck, ''The ACTS Flight System: Cost-Effective Advanced Communications Technology,'' AIAA 10th Communications Satellite Systems Conference, AIAA CP842, Orlando, Fla., pp. 196–201, Mar. 19–22, 1984.

10.15. Lin, S. H., H. L. Bergman, and M. V. Pursley, ''Rain Attenuation on Earth Space Paths—Summary of 10-Year Experiments and Studies,'' *Bell System Technical Journal*, Vol. 59, No. 2, pp. 183–228, Feb. 1980.

10.16. Capsoni, C., and E. Matricciani, ''Orbital and Site Diversity Systems in Rain Environment: Radar-Derived Results,'' Vol. AP-33, No. 5, pp. 517–522, May 1985.

10.17. Engelbrecht, R. S., ''The Effect of Rain on Satellite Communication Above 10 GHz,'' *RCA Review*, Vol. 40, No. 2, pp. 191–229, June 1979.

10.18. Mazur, B., S. Crozier, R. Lyons, and R. Matyas, ''Adaptive Forward Error Correction Techniques in TDMA,'' Sixth International Conference on Digital Satellite Communications, Phoenix, Az., IEEE Cat. No. 83CH1848-1, pp. XII-8 to 15, Sept. 19–23, 1983.

APPENDIX A
ELEVATION ANGLE DEPENDENCE FOR SLANT PATH COMMUNICATIONS LINKS

Very often in radiowave propagation calculations it is necessary to evaluate a path length dependent parameter, such as atmospheric attenuation, path delay, or rain attenuation, as a function of the elevation angle θ of the ground antenna to the satellite. Atmospheric attenuation models are usually developed for the zenith ($\theta = 90°$) direction, and attenuation at other elevation angles must be derived from that value. Rain attenuation modeling, on the other hand, is most often developed for terrestrial paths ($\theta = 0°$) and other values of attenuation for slant paths must be derived from it.

In this appendix the general procedure for determining elevation angle dependence for a slant path is developed.

Consider a horizontally stratified interaction region in the atmosphere of height H above the surface of the spherical earth with an effective radius (including refraction) of R, as shown in Figure A-1. We desire to determine the path length L through the stratified atmosphere, as a function of the elevation angle θ. The value for R is usually taken as 8500 km. H will typically range up to 10–20 km maximum for most parameters of interest for radiowave propagation factors.

The path length L is then found from geometric considerations as

$$L = \frac{2H}{\sqrt{\sin^2\theta + (2H/R)} + \sin\theta} \tag{A-1}$$

This result is valid for all elevation angles from $\theta = 0°$, where $L = \sqrt{2RH}$, to $\theta = 90°$, where $L = H$, under the reasonable assumption that $R >> 2H$.

A further simplification is possible for elevation angles greater than about 10°. For θ between 10° and 90°, $\sin^2\theta$ will range between 0.03 and 1. The maximum value of $2H/R$ will be on the order of 0.0024 for the values discussed above. Therefore for $\theta > 10°$,

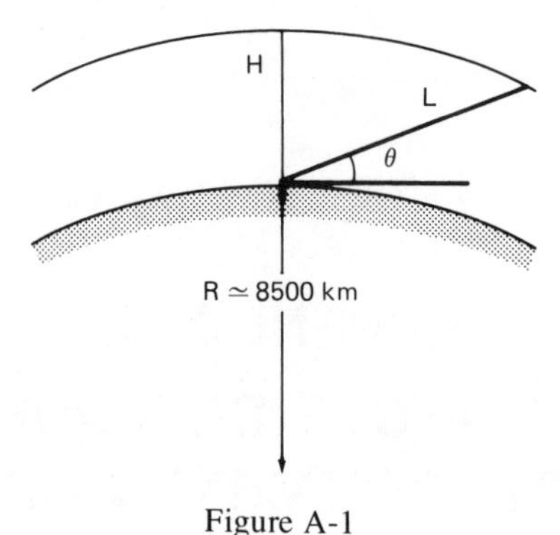

Figure A-1

$$\sin^2 \theta \gg \frac{2H}{R}$$

and

$$L = \frac{H}{\sin \theta} = H \csc \theta \tag{A-2}$$

Equation (A-2) is often referred to as the *cosecant law* for elevation angle dependence on a slanth path. It is only valid, though, for elevation angles above about 10°.

Equations (A-1) and (A-2) can be used to determine the values of path length dependent parameters as a function of elevation angle θ and height H.

For example, to determine the total water vapor attenuation through the atmosphere at an elevation angle θ, $A_w(\theta)$, from the specific attenuation, γ_w(dB/km),

$$A_w(\theta) = L\gamma_w$$

$$= \frac{2H\gamma_w}{\sqrt{\sin^2 \theta + (2H/R)} + \sin \theta}, \quad \text{dB} \tag{A-3}$$

For $\theta > 10°$,

$$A_w(\theta) = \frac{H\gamma_w}{\sin \theta} \tag{A-4}$$

In general, to determine the value of a path length dependent parameter at an elevation angle ϕ, $P(\phi)$, given the parameter at an elevation angle θ, $P(\theta)$,

$$\frac{P(\phi)}{L(\phi)} = \frac{P(\theta)}{L(\theta)}$$

or

$$P(\phi) = \frac{L(\phi)}{L(\theta)} P(\theta)$$

Then,

$$P(\phi) = \frac{\sqrt{\sin^2 \theta + (2H/R)} + \sin \theta}{\sqrt{\sin^2 \phi + (2H/R)} + \sin \phi} P(\theta) \tag{A-5}$$

If both θ and ϕ are greater than 10°,

$$P(\phi) = \frac{\sin \theta}{\sin \phi} P(\theta) \tag{A-6}$$

These results are used in the developments of gaseous attenuation (Chapter 3), rain, cloud and fog attenuation (Chapter 4), and in several other sections throughout the text.

APPENDIX B
INTERPOLATION PROCEDURE FOR ATMOSPHERIC ATTENUATION COEFFICIENTS

Tables 3-1 and 3-2 present listings of frequency dependent coefficients for use in the determination of specific attenuation and total zenith attenuation due to gaseous atmospheric absorption. To determine the coefficients at frequencies other than those given in the tables, the following procedure should be used.

Given the frequency/coefficient pairs f_1/y_1 and f_2/y_2, from either table, where y is $a(f)$, $b(f)$, $c(f)$, or $\alpha(f)$, $\beta(f)$, $\xi(f)$, we desire to determine the coefficient y_0 at frequency f_0 (see Figure B-1).

Note that,

$$\log y_2 = m \log f_2 + b' \tag{B-1}$$

and

$$\log y_1 = m \log f_1 + b' \tag{B-2}$$

Solving for m and b',

$$m = \frac{\log [(y_1/y_2)]}{\log [(f_1/f_2)]} \tag{B-3}$$

$$b' = \log y_2 - m \log f_2 \tag{B-4}$$

The coefficient y_0 at frequency f_0 is then found from

$$\log y_0 = m \log f_0 + b' \tag{B-5}$$

where m and b' are determined from Equations (B-3) and (B-4) above.

For example, to determine the $\beta(f)$ coefficient at 17.5 GHz from Table 3-2,

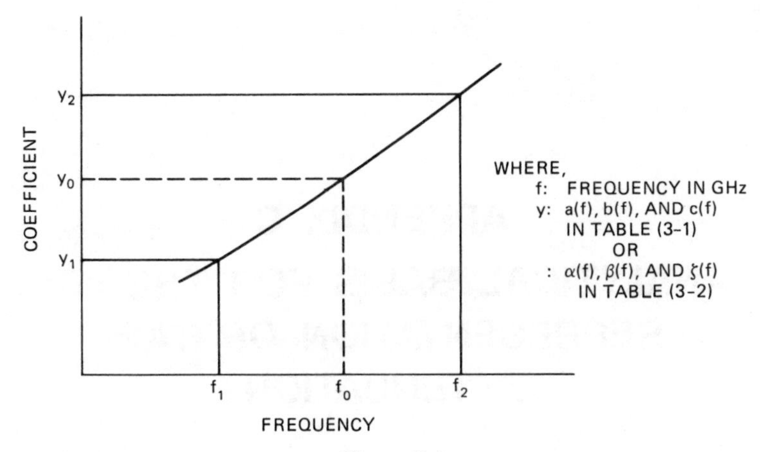

Figure B-1

$$f_1 = 16 \text{ GHz} \qquad \beta_1(f) = 0.00821$$

$$f_2 = 20 \text{ GHz} \qquad \beta_2(f) = 0.0346$$

$$f_0 = 17.5 \text{ GHz} \qquad \beta_0(f) = ???$$

From Equation (B-3),

$$m = \frac{\log [(0.00821/0.0346)]}{\log [(16/20)]} = 6.447$$

From Equation (B-4),

$$\begin{aligned} b' &= \log 0.0346 - 6.447 \log 20 \\ &= -9.848 \end{aligned}$$

Therefore, from Equation (B-5),

$$\log \beta_0(f) = 6.447 \log 17.5 - 9.848 = -1.8341$$

$$\beta_0(f) = 0.01465$$

The $\alpha(f)$ and $\xi(f)$ coefficients for 17.5 GHz can be found in a similar manner.

APPENDIX C
ANALYTICAL BASIS FOR THE aR^b REPRESENTATION OF RAIN ATTENUATION

The specific attenuation produced by rain on a radiowave path was found in Chapter 4 to be well approximated by the expression

$$\alpha \left(\frac{\mathrm{dB}}{\mathrm{km}}\right) = aR^b \tag{C-1}$$

where R is the rain rate in mm/h and a and b are frequency and temperature dependent constants. In this appendix the analytical basis for this relationship is developed from the classical descriptions of rain attenuation on a radiowave path.

Consider the specific attenuation in terms of the attenuation cross-section Q_t and drop size distribution $n(r)$, as given by Equation (4-13)

$$\alpha = 4.343 N_0 \int Q_t(r, \lambda, m)\, e^{-\Lambda r}\, dr \tag{C-2}$$

where [see Eq. (4-12)]

$$\Lambda = cR^{-d}$$

In these equations Q_t is the attenuation cross section, r is the rain drop diameter, λ is the wavelength, m is the refractive index of the water drop, R is the rain rate in mm/h, and N_0, c, and d are empirical constants.

The attenuation cross section is given by [see Equation (4-9)]

$$Q_t = \frac{\lambda^2}{2\pi} \sum_{n=1}^{\infty} (2n + 1) \operatorname{Re}(a_n + b_n) \tag{C-3}$$

In the region where $2\pi r << \lambda$, i.e., where the Rayleigh Scattering condition applies,

$$a_1 \cong \frac{i}{4S}(m^2 - 1)\left(\frac{2\pi r}{\lambda}\right)^5 + \cdots$$

$$b_1 \cong -\frac{2}{3} i \frac{m^2 - 1}{m^2 + 2}\left(\frac{2\pi r}{\lambda}\right)^3 + \frac{1}{15} i \frac{m^4 - 1}{m^2 + 2}\left(\frac{2\pi r}{\lambda}\right)^5 + \cdots$$

Only the first term of b_1 is signfiicant compared to the other terms, therefore

$$Q_t = \frac{\lambda^2}{2\pi} 3 \text{ Re } b_1$$

or

$$Q_t = \frac{8\pi^2}{\lambda} r^3 \text{ Im } \frac{m^2 - 1}{m^2 + 2} \quad \text{(C-4)}$$

where Re indicates "the real part of," and Im "the imaginary part of."

Then, substituting Q_t from Equations (C-4) into Equation (C-2)

$$\alpha = 4.343 N_0 \frac{8\pi^2}{\lambda}\left[\text{Im } \frac{m^2 - 1}{m^2 + 2}\right] \int_0^\infty r^3 e^{-\Lambda r} dr \quad \text{(C-5)}$$

Evaluating the definite integral,

$$\int_0^\infty r^3 e^{-\Lambda r} dr = \frac{6}{\Lambda^4} - \frac{1}{\Lambda} r^3 e^{-\Lambda r} \quad \text{(C-6)}$$

For typical values of rain rate R and drop size r, with the Marshall–Palmer drop size distribution assumed, the second term in the integral will be negligible compared to the first term, as shown by Table C-1.

The integral therefore reduces to the value $6/\Lambda^4$, and Equation (C-5) becomes

$$\alpha = \frac{4.343 N_0 48\pi^2 \text{ Im } \dfrac{m^2 - 1}{m^2 + 2}}{\lambda c^4} R^{4d}$$

or

$$\alpha = aR^b \quad \text{(C-7)}$$

Table C-1. Comparison of Terms in Integral

$$\int_0^\infty r^3 e^{-\Lambda r}\,dr = \underset{\uparrow \atop \text{Term 1}}{\frac{6}{\Lambda^4}} - \underset{\uparrow \atop \text{Term 2}}{\frac{1}{\Lambda} r^3 e^{-\Lambda r}}$$

$\Lambda = 820R^{-.21}$ (Marshall–Palmer Distribution)

R (mm/hr)	r (mm)	Term 1	Term 2
.25	0.25	2.28×10^{-2}	1.74×10^{-16}
	0.75		7.08×10^{-39}
	1.25		4.9×10^{-62}
1	0.25	7.31×10^{-2}	2.38×10^{-13}
	0.75		1.00×10^{-29}
	1.25		7.27×10^{-47}
10	0.25	0.566	1.01×10^{-9}
	0.75		2.84×10^{-19}
	1.25		1.37×10^{-29}
25	0.25	1.09	1.11×10^{-8}
	0.75		2.62×10^{-16}
	1.25		1.06×10^{-24}
50	0.25	1.96	5.27×10^{-8}
	0.75		2.10×10^{-19}
	1.25		1.43×10^{-21}
100	0.25	3.50	2.066×10^{-7}
	0.75		9.48×10^{-13}
	1.25		7.45×10^{-19}
200	0.25	6.27	6.87×10^{-7}
	0.75		2.6×10^{-11}
	1.25		1.69×10^{-16}

where

$$a = \frac{4.343\,N_0 48\pi^2 \,\mathrm{Im}\,\dfrac{m^2-1}{m^2+2}}{\lambda c^4} \tag{C-8}$$

and

$$b = 4d \tag{C-9}$$

APPENDIX D
CRANE GLOBAL RAIN ATTENUATION MODEL CALCULATION PROCEDURE

This appendix presents the step-by-step procedure for the calculation of rain attenuation for an average year by use of the Crane global model, discussed in Chapter 5, Section 5.4.

The input parameters required for the Crane Global Model are:

f: Frequency (GHz)
θ: Elevation angle to satellite (degrees)
G: Ground station elevation, i.e., the height above mean sea level (km)
ϕ: Ground station latitude (degrees)

The mean rain attenuation distribution for an average year is determined as follows:

STEP 1. Obtain the annual rain rate distribution, R_p, for values of p from 2% to 0.001% of an average year, for the location of interest. If this information is not available from local historical data sources, use the appropriate rain rate distribution listed in Table 4-9, as determined from the climate regions given by the maps of Figures 4-8, 4-9, or 4-10.

STEP 2. Determine the 0°C isotherm height $H(p)$ for each percent of the average year p from Figure 4-11. Isotherm heights for $p = 0.001\%$, 0.01%, 0.1%, and 1% can be read directly off of the curves on the figure. Isotherm heights at other values of p can be determined by logarithmic interpolation between the curves.

STEP 3. Calculate the projected surface path length D for *each* p percent of the year desired from the following:

For $\theta \geq 10°$,

$$D = \frac{H(p) - G}{\tan \theta} \tag{D-1}$$

where $H(p)$ are the 0° isotherm heights obtained in Step 2, G is the ground station elevation above mean sea level, and θ is the elevation angle to the satellite.

At elevation angles less than 10°, the curvature of the earth's surface must be accounted for. Therefore, for $\theta < 10°$,

$$D = R \sin^{-1} \cdot \left[\frac{\cos \theta}{H(p) + R} \left(\sqrt{(G + R)^2 \sin^2 \theta + 2R(H(p) - G) + H^2(p) - G^2} - (G + R) \sin \theta \right) \right] \quad \text{(D-2)}$$

where R is the effective radius of the earth, assumed to be 8500 km.

STEP 4. Determine the specific attenuation coefficients a and b at the frequency and polarization of interest. The values given by Table 4-3 are recommended.

STEP 5. Determine the following four empirical constants from R_p for *each* p of interest.

$$d = 3.8 - 0.6 \ln R_p \quad \text{(D-3)}$$

$$x = 2.3 R_p^{-0.17} \quad \text{(D-4)}$$

$$y = 0.026 - 0.03 \ln R_p \quad \text{(D-5)}$$

$$U = \frac{\ln (xe^{Yd})}{d} \quad \text{(D-6)}$$

STEP 6. The mean slant-path rain attenuation at *each* probability of occurrence p is then found as follows:

(a) If $0 < D \leq d$:

$$A(p) = \frac{aR_p^b}{\cos \theta} \left[\frac{e^{UbD} - 1}{Ub} \right] \quad \text{(D-7)}$$

(b) If $d < D \leq 22.5$:

$$A(p) = \frac{aR_p^b}{\cos \theta} \left[\frac{e^{Ubd} - 1}{Ub} - \frac{xe^{Ybd}}{Yb} + \frac{X^b e^{YbD}}{Yb} \right] \quad \text{(D-8)}$$

(c) If $D > 22.5$, calculate $A(p)$ with $D = 22.5$ but use the rain rate R'_p at the value

$$p' = p\left[\frac{22.5}{D}\right] \tag{D-9}$$

instead of R_p.

STEP 7. Estimate the upper and lower bounds of the mean slant-path attenuation (i.e., the standard deviation of measurements about the model) from the following:

Percent of Year	Standard Deviation (%)
1.0	±39
0.1	32
0.01	32
0.001	39

For example, a mean prediction of 12 dB at 0.01% of the year yields an upper/lower bound of ±32% or ±3.84 dB.

APPENDIX E
CCIR RAIN ATTENUATION MODEL CALCULATION PROCEDURE

This appendix presents step-by-step procedures for the calculation of rain attenuation for an average year by application of the CCIR rain attenuation model discussed in Chapter 5, Section 5.5. Three separate methods are included in the CCIR model:

Method I: for maritime climates
Method II: for continental climates and/or for time percentages greater than 0.01%
Method I′: for tropical climates

The CCIR Model determines an annual attenuation distribution at a specified location from as "average year" rain rate distribution. The input parameters required for the model are:

f: frequency (GHz)
θ: elevation angle to the satellite (degrees)
G: Ground station elevation, i.e., the height above mean sea level (km)
ϕ: Ground station latitude (degrees)

The step-by-step procedure for each method follows.

METHOD I. MARITIME CLIMATES

STEP 1. Obtain the rain height h_R from

$$h_R = 5.1 - 2.15 \log [1 + 10^{(\phi - 27/25)}] \tag{E-1}$$

where ϕ is the ground station latitude, in degrees. This equation is shown plotted in Figure E-1 as the curve labeled "Method I."

STEP 2. The slant path length L_S is determined from

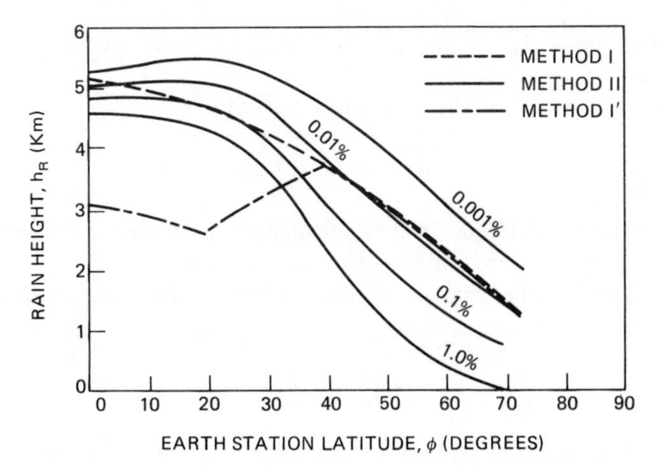

Figure E-1. Rain height as a function of earth station latitude for the CCIR prediction methods.

$$L_S = \frac{2(h_R - G)}{[\sin^2 \theta + (h_R - G/4250)]^{1/2} + \sin \theta} \tag{E-2}$$

where G is the ground station height above mean sea level, in km, and θ is the elevation angle, in degrees. An effective Earth radius of 8500 km is usually assumed.

For $\theta \geq 10$, Equation (E-2) can be approximated by

$$L_S \cong \frac{h_R - G}{\sin \theta} \tag{E-3}$$

STEP 3. The horizontal projection L_G of the slant path length is then found from

$$L_G = L_S \cos \theta \tag{E-4}$$

STEP 4. The reduction factor r_p for 0.01% of the time is calculated by

$$r_p = \frac{90}{90 + 4L_G} \tag{E-5}$$

STEP 5. Obtain the rain rate R_p exceeded for 0.01% of an average year at the location of interest. If this information is not available from local data sources, an estimate can be obtained by determination of the rain climate zone from

Figures 4-13 or 4-14, and the corresponding rain rate value at 0.01% from Table 4-10.

STEP 6. Obtain the specific attenuation α from

$$\alpha = aR_p^b, \quad \text{dB/km} \tag{E-6}$$

where a and b are the frequency dependent constants described by Equation (4-15). The values given by Table 4-3 are recommended.

STEP 7. The rain attenuation exceeded for 0.01% of the average year is then obtained from

$$A_{0.01} = \alpha L_S r_p \tag{E-7}$$

STEP 8. The rain attenuation for other percentages p of the average year is found from Equation (E-7) by

$$A_p = \begin{cases} A_{0.01}\left(\dfrac{p}{0.01}\right)^{-0.33}, & 0.001 \le p \le 0.01 \\ \\ A_{0.01}\left(\dfrac{p}{0.01}\right)^{-0.41}, & 0.01 < p \le 0.1 \end{cases} \tag{E-8}$$

METHOD II. CONTINENTAL CLIMATES

Method II is used for locations in continental climates and/or when time percentages greater than 0.1% are required. It is somewhat more invovled than is Method I, since it uses four rain rate distribution values rather than the single value at 0.01% used in Method I.

STEP 1. Obtain the rain height h_R for the four percentage values 0.001%, 0.01%, 0.1%, and 1% at the ground location of interest from the curves of Figure E-1 labeled "Method II."

STEP 2. Determine the slant-path length L_S from Equations (E-2) or (E-3), as in Method I, Step 2.

STEP 3. Determine the horizontal projection L_G from Equation (E-4) as in Method I, Step 3.

STEP 4. The reduction factor r_p is determined for each of the four time percentages from

$$r_{0.001} = \frac{10}{10 + L_G} \tag{E-9}$$

$$r_{0.01} = \frac{90}{90 + 4L_G} \tag{E-10}$$

$$r_{0.1} = \frac{180}{180 + L_G} \tag{E-11}$$

$$r_{1.0} = 1 \tag{E-12}$$

STEP 5. Obtain the rain rate R_p at the location of interest exceeded for 0.001%, 0.01%, 0.1%, and 1.0% of an average year. If this information is not available from local data sources, use the appropriate rain rate distribution listed in Table 4-9, as determined from the climate regions given by the maps of Figures 4-8, 4-9, or 4-10.

STEP 6. The rain attenuation exceeded for each of the four time percentages is found from

$$A_{0.001} = aR^b_{0.001}L_S r_{0.001} \tag{E-13}$$

$$A_{0.01} = aR^b_{0.01}L_S r_{0.01} \tag{E-14}$$

$$A_{0.1} = aR^b_{0.1}L_S r_{0.1} \tag{E-15}$$

$$A_1 = aR^b_1 L_S r_1 \tag{E-16}$$

where a and b are the frequency dependent constants for specific attenuation described by Equation (4-15).

STEP 7. The rain attenuation for other percentages p between 0.001% and 1.0% is found from Equations (E-13) through (E-16) by the following

$$A_p = \begin{cases} A_{0.001}\left(\dfrac{p}{0.001}\right)^{\log[(A_{0.001}/A_{0.01})]}, & 0.001 < p < 0.01 \\ A_{0.01}\left(\dfrac{p}{0.01}\right)^{\log[(A_{0.01}/A_{0.01})]}, & 0.01 < p < 0.1 \\ A_{0.1}\left(\dfrac{p}{0.1}\right)^{\log[(A_{0.1}/A_1)]}, & 0.1 < p < 1 \end{cases} \tag{E-17}$$

METHOD I′. TROPICAL CLIMATES

Method I was modified by the CCIR IWP 5/2 in May 1982 to improve the prediction for tropical climates where the original Method I was found to overpredict the rain attenuation.

The Method I′ procedure differs from Method I in two ways. First, the rain height h_R calculated in Step 1, Equation (E-1), is modified by a reduction factor ρ_p which is a function of ground station latitude. The modified rain height h'_R is found from

$$h'_R = \rho_p h_R \tag{E-18}$$

where

$$\rho_p = \begin{cases} 0.6, & \phi < 20° \\ 0.6 + 0.02(\phi - 20), & 20° \leq \phi \leq 40° \\ 1.0, & \phi > 40° \end{cases} \tag{E-19}$$

The modified rain height is shown plotted in Figure E-1 as the curve labeled "Method I′."

Method I Steps 2 through 7 are unchanged for Method I′.

The second modification occurs in Step 8, the calculation of rain attenuation for other percentages of the year. Method I′ includes an additional equation for calculating rain attenuation for percentages from 0.1 to 1.0%. Equation (E-8) is extended to

$$A_p = \begin{cases} A_{0.01}\left(\dfrac{p}{0.01}\right)^{-0.33}, & 0.001 \leq p \leq 0.01 \\ A_{0.01}\left(\dfrac{p}{0.01}\right)^{-0.91}, & 0.01 < p \leq 0.1 \\ 1.3A_{0.01}\left(\dfrac{p}{0.01}\right)^{-0.5}, & 0.1 < p \leq 1 \end{cases} \tag{E-20}$$

Method I′ provides predictions which are identical to Method I for locations above 40° latitude, and provides predictions which are reduced by up to about 40% for locations below 20° latitude.

Method I′ was adapted by CCIR Study Group V at it's 1983 Interim Meetings in Geneva as the sole CCIR Prediction Model to be used for attenuation prediction calculations (CCIR Doc. 5/101, 27 July 1983).

APPENDIX F
CCIR TROPOSPHERIC SCINTILLATION MODEL PROCEDURE

In this appendix the detailed step by step procedure of the CCIR tropospheric scintillation model introduced in Chapter 8 is presented. A thin turbulent layer at an average height of 1 km is assumed, and empirical approximations to the determination of amplitude fluctuations from turbulence theory are employed in the model.

The model determines the standard deviation of the log of the received power, which is a measure of the root mean square (r.m.s.) amplitude scintillation of a radiowave transmitted on a satellite path. The model is applicable at any elevation angle, and has shown good agreement with measurements at frequencies up to 30 GHz.

REQUIRED INPUT PARAMETERS

Antenna diameter D, in meters
Operating frequency f, in GHz
Elevation angle θ, in degrees

STEP 1. Determine L, the slant path distance to the horizontal thin turbulent layer, from

$$L = [\sqrt{0.017 + 72.25 \sin^2 \theta} - 8.5 \sin \theta] \times 10^6 \qquad \text{(F-1)}$$

STEP 2. Determine the parameter Z from

$$Z = 0.685 \frac{D}{\sqrt{\frac{L}{f}}} \qquad \text{(F-2)}$$

STEP 3. Determine the antenna aperture averaging factor, $G(z)$, from

$$G(z) = \begin{cases} 1.0 - 1.4z, & 0 < z < 0.5 \\ 0.5 - 0.4z, & 0.5 < z < 1 \\ 0.1, & 1 < z \end{cases} \tag{F-3}$$

STEP 4. The r.m.s. amplitude scintillation, expressed as σ_x, the standard deviation of the log of the received power, is then given by

$$\sigma_x(\text{dB}) = 0.025 f^{7/12} [\csc \theta]^{0.85} [G(z)]^{1/2} \tag{F-4}$$

SAMPLE CALCULATION

Let

$$D = 37 \text{ m}$$
$$f = 7.3 \text{ GHz}$$
$$\theta = 10 \text{ degrees}$$

STEP 1.

$$L = [\sqrt{0.017 + 72.25 (\sin^2 10)} - 8.5 \sin 10] \times 10^6$$
$$= 5747 \text{ m}$$

STEP 2.

$$Z = 0.685 \frac{37}{\sqrt{\dfrac{5747}{7.3}}} = 0.903$$

STEP 3.

$$G(z) = 0.5 - 0.4(0.903) = 0.139$$

STEP 4.

$$\sigma_x = 0.025(7.3)^{7/12}[\csc 10]^{0.85}[0.139]^{1/2}$$
$$= 0.13 \text{ dB}$$

It should be emphasized that the model is based on a statistical description of the propagation path. The r.m.s. amplitude scintillation calculated by the model should be expected over a long term averaging period (several months). Instantaneous, short term fluctuations could exceed this value by several orders of magnitude.

INDEX

INDEX